Geological Features and Mysteries of THE PLANET EARTH

Other Books by Dr. N.W. Gokhale

- Manual of Geological Maps
- Fundamentals of Sedimentary Rocks
- A Manual of Problems in Structural Geology
- Exercises on Geological Maps & DIP-Strike Problems
- Theory of Structural Geology
- A Guide to Field Geology

Geological Features and Mysteries of THE PLANET EARTH

Dr. N.W. Gokhale
M.Sc., Ph.D. (Budapest), F.M.S. (Mysore), F.I.S.E.S. (Pune)
Ex-Professor of Geology, Department of Studies in Geology,
Karnataka University, Dharwad (Karnataka)

CBS

CBS PUBLISHERS & DISTRIBUTORS PVT. LTD.
New Delhi • Bengaluru • Chennai • Kochi • Mumbai • Pune

ISBN : 81-239-1288-9

First Edition : 2006
Reprint : 2010, 2014

Published by Satish Kumar Jain for
CBS Publishers & Distributors Pvt. Ltd.,
CBS Plaza, 4819/XI Prahlad Street, 24 Ansari Road, Daryaganj,
New Delhi - 110002, India. • Website: www.cbspd.com
e-mail: delhi@cbspd.com, cbspubs@airtelmail.in
Ph.: 23289259, 23266861, 23266867 • Fax: 011-23243014

Branches:

- ***Bengaluru:*** Seema House, 2975, 17th Cross, K.R. Road, Bansankari 2nd Stage, Bengaluru - 560070
  • Ph.: +91-80-26771678/79 • Fax: +91-80-26771680
  • E-mail: cbsbng@gmail.com, bangalore@cbspd.com
- ***Pune:*** Bhuruk Prestige, Sr. No. 52/12/2+1+3/2, Narhe, Haveli (Near Katraj-Dehu Road By-pass), Pune - 411041
  • Ph.: +91-20-64704058/59, 020-32392277 • E-mail: pune@cbspd.com
- ***Kochi:*** 36/14, Kalluvilakam, Lissie Hospital Road, Kochi - 682018, Kerala • Ph.: +91-484-4059061-65
  • Fax: +91-484-4059065 • E-mail: cochin@cbspd.com
- ***Chennai:*** 20, West Park Road, Shenoy Nagar, Chennai - 600030
  Ph.: +91-44-26260666, 26208620 • Fax: +91-44-42032115
  • E-mail: chennai@cbspd.com
- ***Mumbai:*** 83-C, Ist Floor, Dr. E. Moses Road, Worli, Mumbai-400 01
  Maharashtra Ph.: +91-9833017933, 022-24902340/24902341
  • E-mail: mumbai@cbspd.com

Printed at :
J.S. Offset Printers, Delhi

FOREWORD

FOREWORD

The book entitled "Geological Features and Mysteries of the Planet Earth" by Prof. N.W. Gokhale deals with the various aspects of earth such as nature of earth's crust, mantle and core, continents, oceans, earthquakes with special reference to Plate Tectonics, crystallization of magma, and causes of salinity of sea water. It also scrutinizes the position of earth with respect to the cosmos. The subject matter of the book that has been described in nine chapters includes geomorphology, igneous petrology, economic geology, groundwater and some aspects of cosmic geology.

Origin of the universe, inclination of the axis of the earth is informative and interesting especially to an inquisitive reader for understanding some fundamental questions, which have not been so far completely understood. A number of intriguing problems that have been discussed in the book include : (a) the disproportionate extent of hydrosphere and land; (b) superabundance of nitrogen and oxygen (amongst gases) and that of Si, Fe (amongst other elements); (c) inclination of Earth's axis; (d) salinity of sea water; (e) age of water; and (f) shapes and positions of various continents.

Each one of the nine chapters has revealed thought-provoking issues that were apparently overlooked earlier. Discussion on the origin of continents and oceans with reference to Plate Tectonics is quite interesting. The inadequacies of the Plate Tectonics theory have been pointed out. The chapter on the earthquakes has reasoned out as to why these should be deemed as constructive and not destructive events. Crystallization of silicate minerals from Magma and formation of elements have been discussed. The discussion pertaining to continental drift, seafloor spreading, convection current hypothesis in the light of upto date data, is quite thought-provoking.

Topics described in the book reflect that the author has gone into the depth and has given new dimensions to many complex aspects of the earth. The author has resorted to philosophy and mysticism in order to explain some aspects of the unresolved problems of the earth sciences.

January 2005 **(L.N. Gupta)**

PREFACE

PREFACE

The ease with which scientists are investigating planets in the Universe, they are, however, handicapped in not being able to reach the interior parts of the Earth, and collect data directly. The geological history of the Earth is composed from piecemeal data mainly collected from the crust. A careful scrutiny discloses shortcomings in the modus operandi tendered to account for some of the features from petrology, geomorphology, earthquakes, Convection Current Hypothesis, plate tectonics, and so on. Realisation of the shortcomings has prompted me to describe them in this book. Geological episodes are required to be visualised and depicted in "three dimensional diagrams". Even at the end of the 20th century, convincing explanation is not forthcoming at least for some features of the Earth. The book draws the attention of the geo-scientists to them. Some misconceptions have been rectified.

Consultation and discussion held with professors of Chemistry, Physics, Botany, Geology, Geography, as well as with geoscientists from the Department of Mines and Geology, Karnataka State, has immensely helped me to draft the text of the book. I am indebted to Dr. N. Umakanth, retired Professor of Nuclear Physics, Department of Studies in Physics, Karnatak University, Dharwad, late Dr. V.M. Korwar, Professor of Physics, Department of Studies in Physics, Karnatak University, Dharwad, Dr. A.M. Karguppikar, Professor of Physics, Department of Studies in Physics, Karnatak University, Dharwad, Dr. N.C.

Vijayaraj, retired Professor of Geography, Department of Studies in Geography, Karnatak University, Dharwad, Dr. S.C. Puranik, Professor of Geology, Department of Studies in Geology, Karnatak University, Dharwad, Dr. B.P. Waghmare, Professor and Chairman of the Department of Studies in Geology, Karnatak University, Dharwad, Dr. V.G. Gokhale, retired Professor and Head of the Department of Chemistry, and Principal of Arts, Commerce and Science College, J.S.S., Vidyagiri, Dharwad, Dr. G.S. Pujar, Reader in Geology, Department of Geology, Karnatak Science College, Dharwad, Dr. R.S. Godbole, retired Professor and Head of the Department of Physics, Kittle Arts and Science College, Dharwad, Dr. P.S. Agarkar, retired Professor and Head of the P.G. Department of Geology, Science College, Jabalpur, Dr. G.V. Hedge, Geologist, Department of Mines and Geology, Karnatak State, Dr. U.K. Kulkarni, retired Professor of Botany, Department of Botany, Shivaji University, Kolhapur, who have obliged me by sparing their time to hold useful discussions with them. I am thankful to Dr. D. Muralidharan, Scientist, National Geophysical Research Institute, Hyderabad, Andhra Pradesh, for kindly sparing two latest seismograms and that too from the vicinity of India, that have enriched the content of chapter dealing with the earthquakes.

I am specifically grateful to Dr. N. Umakanth, Dr. R.S. Godbole, Dr. B.P. Waghamre and Dr. G.V. Hegde, for their useful suggestions and critically and carefully going through some of the chapters. Last but not the least, I place on record my grateful thanks to Dr. L.N. Gupta, Professor of Geology, Centre of Advanced Studies in Geology, Punjab University, Chandigarh, for kindly accepting my request to write the Foreword for this book. I am thankful to him for sparing time from his busy schedule and drafting excellent Foreword, which will arouse interest in the minds of the students and geologists desirous of learning more about the Earth.

N.W. Gokhale

CONTENTS

CONTENTS

CHAPTER

1

INTRODUCTION

BACKGROUND

Man is trying to understand the various aspects of the Earth, its composition, the atmosphere surrounding it, its very origin, and so on. Since time immemorial, explanations have been tendered on the basis of data collected from various sources. The mechanisms suggested being not convincing ones, those have not gained scientific acceptability. There are many lacunae, particularly in geology. The scientific approach consists of conducting experiments and comparing the results thus obtained with the features encountered in the Earth. But some of the phenomena (events) cannot be verified by conducting experiments within the four walls of any laboratory, howsoever large it might be. The events are, therefore, deemed as "remotely controlled phenomena", thus taking the shelter of philosophical explanation. Do the volcanoes and the earthquakes occur to destroy the life on the Earth? Why plant or animal life alone was not initiated on the Earth? There are many other features that are not explained on convincing scientific basis. Uncertainty prevails in the scientific world, at least in the domain of geology. Above noted and similar other features are generally attributed to the power of *God*, the *Creator* of the *Universe*.

It is interesting to note that the higher degree in the faculty of science is a Ph.D., *Doctor of Philosophy*. It is not termed as *Doctor of Science*, probably because, the scientific pursuits reach a stage when the reasoning and arguments come to a halt. This situation also could be described as reaching a stage of "that far, no further". Features from branches of geology have been analysed and discussed in this book.

GEOMORPHOLOGY

The distribution of land and the oceans has not been explained as to why it is 25 to 75 per cent, respectively. It as well could have been 50 : 50 or 75 : 25, respectively for the land-masses and the oceans. Then, the oceanic body is most extensive between the western coasts of America and the eastern coasts of Russia and China. The shapes of the continents are of interest too. The North and the South America, Greenland and Africa, are very broad towards their northern border, while these taper off into a tail at their southern end. Why is it so? Does it reflect the process and the act of evolution of the continents? The surficial extent (length and breadth) of the continents when considered, Asia, China, Russia and Europe, together constitute one *contiguous landmass*, and it is by far the most extensive land surface.

Generally, the continents are looked in two dimensions, namely the length and the breadth. But this outlook, unknowingly leads one to consider the continents to be independent and separate entities on the surface of the Earth. However, the continents are all connected to each other in the form of the continental slopes, and the bottom surface (land) of the oceans. In other words it means that the surface of the Earth is a contiguous feature. The oceanic stretches are merely huge and gigantic depressions, some of them being very deep, like that of the Pacific ocean. What caused these depressions? Do these reflect changes that are taking place in the interior parts of the Earth?

Regarding the disposition of the continents and the oceans, two views are in vogue : (a) these are permanent features and not to have changed their places, (b) the continents to have drifted from place to place. Both views have shortcomings and loopholes with them. Therefore, uncertainty prevails about the modus operandi of the disposition of the continents and the oceans on the surface of the Earth. The proponents of the continental drift have considered the shapes of South America, South Africa, India and Australia and fitted-up to constitute one landmass, and called it as the Gondwanaland. Such a fit has been prompted from the eastern bulge of South America, and the angle of reentrant of the western coast of Africa. But in so doing, it has been completely forgotten that there are the continental slopes in between the two continents. Therefore, the surficial (two dimensional) shapes are misleading. As noted earlier any landmass (continent) is a three dimensional body, and this aspect is unfortunately not taken into account while propounding the hypotheses regarding the drifting of the continents. Are the continents made up of siaslic material, resting over the sima occurring beneath it? What is the nature of the contact between the sial and the sima? Unless it is weak, the continents can not be dragged or drifted from place to place. The motive force that brought about the movement (drifting or wandering) of the continents, is not described or connected to any geological episode like, the formation of rocks in the crust of the Earth. Convection currents can not be considered or designated as constituting a geological episode, or as an event, because it is a mechanism and not a product. In the first place, why should the continents drift at all? Does it reflect instability or an activity of reconstruction of the crustal part of the Earth? The hypothesis of continental drift is called into play to account for the similarity of certain geological features of South America, Africa, India and Australia. The process of drifting appears to be a mere guess. The Gondwanaland broke apart and then only the drifting came into operation. The act of breaking apart is indicative of tension at that place. Then the question crops up as to what caused the tension? When two continents move towards each other, what happens to the continental slopes, and to the enormous volume of oceanic water that existed in between the two continents?

PLATE TECTONICS

This has been envisaged to account for the numerous geological episodes like the mountain building, the development of mineralised zones in a specific linear manner, generation of magmas, cause of earthquakes, and so on. Plate is a new term for the continents, because the plates that have been enumerated, those correspond to the present-day continents. The plates, however, have not been presented in a three-dimensional form, nor the initial nature of contact with each other, has been described. Do the present-day coastlines mark the upper part of the plates? If so, the plates are wide apart from one another. But what about the nature of the bottom part of these plates? Are those in contact with each other, or those too are gaping apart? It is said that the plates either move (drift) towards or away from one another. Does that mean that the plates, to begin with, were not in contact with one another? If so, how and when were the plates (continents) formed? What compelled them to drift, either towards or away from each other?

COMPOSITION OF EARTH

The Earth is considered to be made up of three shells, based on the observations of increase in the density of rocks, with depth. The three shells are (i) the Crust (sial, density 2.8), (ii) the Sima (density 3.6), and (iii) the Core (density 8.00). The sial makes up the continents, and the sima forms the bottom part of the oceans. The core is considered to be made up of iron and nickel, and is therefore designated as the "*Nife*" core. The extent of the three shells is different. By and large the sial, the sima and the core, respectively are 120, 2700 and 3400 km in thickness.

Literature, however, reveals considerable variation in these values of thickness. The surface of contact of the bottom part of the continents (sial), with the top part of the sima, is of importance, because this will have a bearing upon the drifting of the continents, which in turn will have an impact upon the mechanism of Plate Tectonics.

ORIGIN OF EARTH, ATMOSPHERE AND OCEANS

This issue is still being debated as to whether the Earth cooled from a gaseous state, or it was formed from the accretion of innumerable meteorites. These two mechanisms have been described as the "Big Bang", and the "Planetesimal" hypotheses, respectively. Let us not go still backwards and inquire as to how the material of the Earth was acquired, because, then the frontiers of geology will be crossed, and the discussion will enter the realm of "Cosmic Geology". Starting stage of the Earth is that it was in a molten state, and then the molten matter commenced cooling and gave rise to a solidified body. This is one concept. In the second concept (the Planetesimal hypothesis), the situation is just the opposite one. The Earth was initially in a solid state, and was much smaller in size than what it is today. With the impact of numerous meteorites on it, it grew in size. The core was then rendered molten, and eventually the entire body became a molten mass. It underwent subsequent cooling and resulted into a solidified body — the Earth of the present-day.

The process of consolidation of the molten Earth, probably commenced with the gases escaping into the outer space to form the atmosphere composed of nitrogen and oxygen. The water vapour appears to have ultimately condensed to form the lakes, the seas and the oceans. Prior to the formation of the atmosphere, the outermost part of the molten body must have chilled to form a glassy solidified crust. Below such a shell, the hot molten matter (magma) still persisted, which in the course of time (billions of years) consolidated to form the three shells (Sial, Sima and Core) of the Earth. *Nife* (nickel and iron) settled to the core parts because of higher value of density, while the silicates and the sulphide minerals occupied the peripheral shells of the Earth. It is of interest to observe that nickel and iron are not uniformly distributed throughout the different parts of the Earth, as are the other elements, Na, K, Mg, Ca, Pb, Cr, and so on. Gold is the heaviest element, yet surprisingly, it has not settled at the core, but is occurring in the crustal rocks. This is strange feature indeed!

The building activity of the Earth continued even after the consolidation of the primordial molten matter. These activities, however, have taken place in the shells of sial and sima. Thus occurred the volcanic erruptions of the Deccan Traps of India, the Laki Lava flows of Iceland, the Bushveld Igneous Complex, the Skaerggard Gabbro Intrusion of Greenland, and so on. These events indicate that the building up (construction) of the Earth is a continuous process, which means that even today such activities might be in operation!

CRYSTALLISATION OF MAGMA

About the process of crystallisation of a magma, there is some ambiguity. It is attributed to reaching the stage of saturation of the minerals present in the magma. How far is this interpretation tenable? Does it mean that the silicate minerals are dissolved in a solvent? If so, what is that solvent, and what happens to it after the completion of the process of crystallisation? In actuality, however, magma is not a solution, but is a melt. Therefore, the principle of saturation is not extendable to the crystallisation of magma. In actuality, the individual minerals crystallise from the magma, as soon as their freezing point (melting point) is attained by the magma, and not because the minerals had reached the stage of "saturation". The textures of the igneous rocks indicate that the first minerals to crystallise are those having highest melting point (temperature). Therefore, the melting points (freezing points) and not the stage of saturation of minerals govern the process of crystallisation of magma. It is more reasonable and appropriate to describe the "curve of saturation" as a "curve of melting/freezing point"

SALINITY OF OCEANS

Let us now consider the salty nature of the oceanic waters. The contents of Ca and Mg are considerably more than that of Na in the crustal rocks, but in the oceanic waters, this situation is reversed. There are many "land locked" lakes like the Manasarovar, Lonar (both from India), lake Balaton (Hungary), lake Superior (USA), lake Victoria (Africa), and so on, which are sweet water bodies. Were the oceans and the seas salty when formed, or these were also sweet water bodies? When were the seas and the oceans formed?

GROUNDWATER

It is of interest to observe that the groundwater, bore well and dug well water of some regions is brackish. Is the degree of brackishness (salinity) of such waters similar to that of the seas and the oceans? What is the depth in which the groundwater can reach? Is the permeability of the rocks, the sole governing factor, and/or the rise of temperature with depth, also controls the depth to which the groundwater could seep? What is the mode of occurrence of the groundwater? Does it occur in the form of reservoirs or pools? What is the nature of the groundwater below the oceans and the seas? Hot springs though not frequent, they do exist. Is the source of heat for such springs the ancient volcanoes, the last one being of Cretaceous age (140 m.y.) as far as India is concerned? Or, do the hot springs indicate that igneous activity (lava or magma) is taking place presently at depth?

ECONOMIC GEOLOGY

The industrial and numerous economic minerals are almost restricted and are concentrated in the rocks belonging to the pre-Cambrian period. Almost all of the metalliferous deposits (gold, iron, manganese, chromite, lead, ilmenite, and so on) were formed during the pre-Cambrian period. What was unusual during that period which gave risk to rich deposits of the said minerals all over the world? Likewise, the Eocene period (65 m.y.) has resulted in the formation of oil, the Carboniferous period (350 m.y.) gave birth to coal. Notwithstanding any other reason, could these selective events of formation of economic mineral deposits (iron, gold, coal, oil) be considered as controlled only by time (pre-Cambrian, Carboniferous, Cretaceous)?

PUZZLING FEATURES

(a) The mineral wealth, (b) water bodies (oceans and seas) being more in extent than the landmasses, (c) envelope of atmosphere made up of oxygen and nitrogen, (d) presence of ozone layer, (e) Earth revolving on its axis causing day and night, (f) axis of the Earth at angle of 23½° with respect to the plane of rotation around the Sun causing the different seasons, and many other features possessed by the Earth, are thought-provoking, that are beyond the reach of scientific explanation, at least at present. Consider the feature of presence of both animal and plant life on the Earth. Why not only animal or the plant life came into existence? Earth is having all things that are essential for sustaining human life, like water, oxygen, coal, ore minerals, crude oil, and so on.

It is not known whether the different features possessed by the Earth are also present on the other planets rotating around the Sun. Therefore, there are the features of the Earth to be construed as "accidental" ones, or to be remotely monitored by a "*Creator*"? All the said features of the Earth point to some Supernatural Power at work, that has catered to the needs of the animal and the plant life that was to emerge on the Earth!

CHAPTER

2

COSMOS, UNIVERSE AND THE EARTH

INTRODUCTION

Even before probing into the internal constitution of the Earth, it is very interesting to know as to how the Earth came into existence. It takes us into the empty (?) space and into the Universe. It has been established that the Sun and the distant stars are burning masses of hydrogen. Therefore, it may be said that there are numerous big and small burning masses of hydrogen in the empty space. Why are they burning and wherefrom the element hydrogen came into the space, is a topic that falls into the domain of mysticism. Therefore, hydrogen is the first entrant on the "stage of space", the temperature necessary for its burning being 10^{12} million degrees Kelvin. How such high temperature is generated, is not yet clearly understood. The element hydrogen itself is formed from "Protons, Electrons and Radiations". Who created all these? It is a million or even a billion dollar question. It is certainly necessary to set a limit to going backwards into the Universe, otherwise the discussion goes on becoming more and more mystique and merging into the domain of philosophy. Fowler (1967) writes, "men of strong convictions, religious or scientific, have the courage to deal with the problem of creation". The scientific probe reaches the stage of "that far, no further", leading to mysticism and knocking at the door of, call it "God or some Supernatural Power".

ORIGIN AND COMPOSITION OF UNIVERSE AND THE PLANETS

Earth is one of the innumerable bodies composing the Universe, which is also described as the Cosmos. What is the dimension of the Universe and its composition? Is there only one or are there several Universes? In this context various models have been described below.

Cosmos or Universe is an empty space, which according to some scientists is expanding, while some others consider it to be contracting. The space is held to be made up of *Ether*, while some others consider it to be constituted by hydrogen. What is the expance of space? This aspect might be grasped if the statement "The space has accommodated 10^{20} stars in it, the distance separating two stars from one another being millions of light years", could be understood. The distance between the Sun and the Earth is about 1,70,000 light years, and one light year corresponds to 10^{13} kilometers.

G.L. de Buffon in 1749 suggested that the different planets originate through detachment of portions of the Sun by collision or near-collision with a great comet. According to George Gamow (1961), the cosmos started from a core which exploded in a primordial "Big Bang", some 5 b.y. ago. The fleeing matter from the core, formed the stars, the planets and the galaxies. New theory of synthesis of elements championed by Fred Hoyle (as quoted by Fowler, 1967), starts with a universe consisting of a cold, dilute and turbulent gas of hydrogen atoms. By gravitational force, part of the gas condenses into stars. As a star contracts under gravitational force, its interior grows very dense and hot. When the central temperature reaches about 5 million degrees Kelvin, protons fuse to from deuterons. The deuterons in turn form helium. The net result of the proton-proton chain (Fig. 1) is the conversion of four atoms of hydrogen into one atom of helium. The star thus develops a hot core of helium. The fusion of hydrogen to helium is the source of energy of most of the stars.

Fowler (1967, p. 11) turning to mysticism wrote, "there is food for philosophical thought. The heavy elements, of which our Solar System has its full share, took a long time to produce - probably one to two billion years. Thus the particular part of the universe we inhabit is not the oldest thing in it; many cosmic events preceded the formation of the Earth. The oldest stars in our galaxy are estimated to have an age of 6.5 b.y., while analyses of meteorites indicate that the Solar System is no more than 4.5 b.y. old". According to Hawking (1989), the Universe is deemed as a "Hot Big Bang model" of Friedman. About 100 seconds after the Big Bang, the temperature falls to 10-12 million degrees Kelvin which is the temperature inside the hottest stars. At this temperature, protons and neutrons combine to form "deuterium" (heavy hydrogen). Deuterium then changes to helium. Fowler (1984) has noted that hydrogen constitutes as much as 93%, while helium is a mere 7% in the Universe. According to another estimate, the contents of these elements is 75 and 25 per cent, respectively. What is pertinent to note is that helium itself is produced from hydrogen, and these two elements give rise to the various elements that are encountered in the Universe.

The formation of elements requires burning of hydrogen and helium. It is indeed strange to observe that the fuel for the process of burning, is hydrogen. The burning of hydrogen is

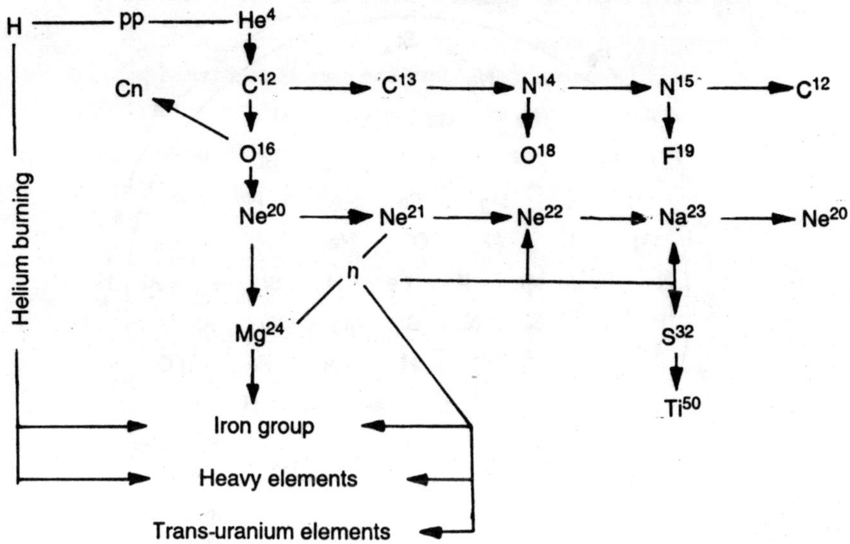

pp = proton - proton reaction
Cn = Carbon - Nitrogen cycle
n = Neutrons liberated
H = Hydrogen, Ne = Neon
He = Helium, C = Carbon
O = Oxygen, S = Sulphur
Ti = Titanium
N = Nitrogen
Na = Sodium
Mg = Magnesium

Fig. 1. Scheme of formation of elements (after Fowler, W.A., 1984).

a "self-sustained process". George Gamow (op. cit.) opines that the formation of several elements lasted just a few minutes, while according to Hawking (op. cit.), within a few hours of the Big Bang, production of helium and the other elements stops. Krane (1988, p. 764) has noted that the formation of several elements lasts over a period of 70,000 years. According to Dott and Batten (1976), "to explain the planet's evolution as well as the mechanism and history of differentiation of Solid Earth, Ocean and Atmosphere, is the greatest challenge for the geologists. The Earth started with a homogeneous, accredited, cold protoplanet about 5 b.y. ago, and it later contracted and healed internally (through radioactivity). Density differentiation produced a very dense core (about 4.5 b.y. ago), less dense crust (about 3.7 b.y. ago), and least dense sea water and atmosphere (about 2.5 b.y. ago). The Earth is still undergoing chemical evolution". Dott and Batten (op. cit.) have presented the process of differentiation in two figures, which have been reproduced here as Figs. 2A and 2B

The burning of hydrogen and helium takes place only in stars like the Sun. No sooner material sparks off from the stars, the burning process ceases immediately. The feature to be noted is that all the elements are formed only during the act of burning. Therefore, it may be said that all the elements encountered in the Earth, must have been formed before it sprang off from the Sun.

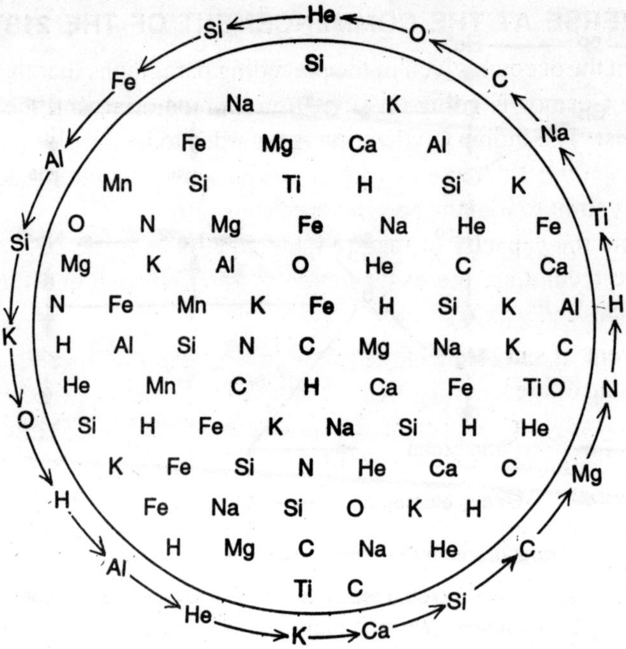

Fig. 2A. Initial accretion. Spinning gaseous mass not differentiated protoplanet. Arrows stand for spinning action.

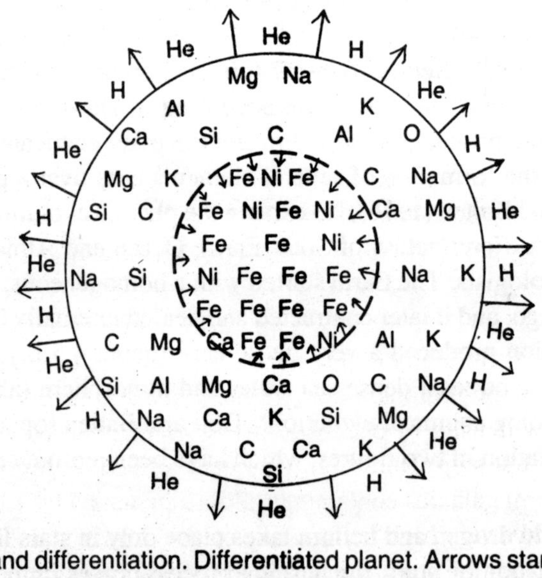

Fig. 2B. Contraction and differentiation. Differentiated planet. Arrows stand for movement of the elements.

VIEW OF UNIVERSE AT THE COMMENCEMENT OF THE 21ST CENTURY

It is apparent from the account given in the preceding paragraphs that the exploration of the Universe and the Cosmos is going on since time immemorial, and the views differ from scientist to scientist. The Indian Civilization is regarded to be the oldest one, and the Vedas (scriptures) describe the Universe as "neither to have any beginning nor an end. It is not possible to set any limit to it in the parameter of *time* or *spatial* extent. It is a dimensionless entity. It is beyond the capacity of man to understand its vastness". Despite the advancement in scientific technique, the exact nature of the Universe and the Cosmos has still remained unresolved.

With the advent of satellite technology, probing deep into the *space* is now possible, and to prepare a map of the Universe and the Cosmos. A Satellite called WMAP (Wilkinson Microwave Anisotropy Probe) was launched on June 30, 2001, that orbitted the Earth, and recorded cosmic emanations from a point on the other side of the Moon. Dr. David Spergel, a Princeton University Astrophysicist, and a member of the WMAP team says, "I think there is no longer any credible alternative to the *Big Bang*". The Universe according to WMAP is 13.7 b.y. old, plus or minus 1 per cent. It is geometrically "flat", in accordance with the simplest solutions of Albert Einstein's equations which equate gravity with the bending of space time. By weight it is 4 per cent atoms, 23 per cent dark matter - presumably as yet undiscovered elementary particles left from the Big Bang - and 73 per cent "dark energy". According to WMAP scientists, their data was beginning to shed light on a theory of what had been going on during the Big Bang. The theory known as "Inflammation", hypothesizes that the Universe underwent an enormous growth spurt during the first trillionth of trillianth of a second of time, under the influence of brief but powerful antigravitational field, that permeated the space (New York Times, 12th February 2003. Reported by The Hindu, India's national newspaper, 13.2.2003). To understand the Universe and the Cosmos is certainly an intellectual challenge to the mankind.

FORMATION OF ELEMENTS DURING BIG BANG

Though there are as many as 10^{20} stars in the Universe, hydrogen gets burnt appreciably only in some stars, which explode. A lot of hydrogen in the Universe is unburnt. Elements composing the planets inclusive of Earth, are built up in a systematic manner. As noted earlier, the Universe is composed of hydrogen, it being the first element. Very dense matter of 10^6 tons/cc explodes and protons, electrons and radiations expand into space and get too thinly spread to a low density of one atom per cc. This is designated as the "Big Bang Stage", and it is the first step. Due to gravitational attraction thinly spread matter starts coming together to form galaxies containing millions of stars. Due to high pressure, a density of 10^6 tons/cc, and a very high temperature of 10-12 million degrees Kelvin is created in the central part. Fusion of protons leads to the liberation of high energy. The Carbon-Nitrogen cycle also produces energy. Through the thermonuclear fusion, several elements are built up. Successive capture of neutrons and emission of electrons, high elements upto uranium are produced. The elemental composition of the Universe is built-up in this fashion. Fowler (1967) has presented the formation of the elements in a schematic manner, and the same has been set out in Fig. 1.

The scheme of Fowler indicates that the proton-proton reaction (fusion) produces helium from hydrogen. From helium all other elements are formed. Formation of elements Si, Al, Na, Ca, Mg releases heat, and these elements, therefore, could change into one another, through reversible reactions. Fe, Ti, Co and the other heavy elements are formed with absorption of heat. These elements are not convertible to other elements because the reaction is not a reversible one. Do the reversible and the non-reversible thermonuclear reactions (fusions) control the quality and the quantity of the several elements encountered in the Universe? In this context, information given by Krane (1988, pp. 770-771) is extremely useful and invaluable. Some of the reactions described by him are given below :

$Ne^{20} + He^{4} \rightarrow Mg^{24} + \gamma$, $O^{16} + O^{16} \rightarrow Si^{28} + He^{4}$

$Si^{28} + \gamma \rightarrow Mg^{24} + He^{4}$, $Si^{28} + He^{4} \rightarrow S^{32} + \gamma$, $Si^{28} + Si^{28} \rightarrow Ni^{56}$

$N^{14} \rightarrow O^{18} \rightarrow Ne^{22} \rightarrow Mg^{26}$ (γ stands for energy released).

Krane writes (p. 771) "The end products of chains of such reactions are the mass 56 nuclei (Ni^{56}, Co^{56}, Fe^{56}). At this point there is no longer energy released in the capture reactions, and the process halted". It is also necessary to know the quantity of each element composing the Universe. Krane (op. cit.) has presented the relative abundances of the elements in the form of a graph, and that has been reproduced as Fig. 3.

Scrutiny of Fig. 3 reveals that oxygen $10^{7.5}$ is the most abundant element, least being beryllium 10^{0}. Rest of the elements all taken together have a value of $10^{2.8}$. The several elements have been presented below in the decreasing order of abundance.

O, C, N, Ne, (Mg, Si, Fe), Ar, S, (Al, Ca, Na), Ni, Cr, P, (K, Co), (Cl, Ti, Zn), Mn, V, Li, Sc, B and Be.

Earth is a planet, its elemental composition, therefore, could be expected to be similar to that described by Krane (op. cit.) for the other planets. Major elements building up the minerals constituting the crustal rocks, are Si, Al, Na, Ca, K, Fe and Mg.Whether the volume of the rocks produced by the minerals is in agreement with the value of abundances of the elements noted by Krane (op. cit.) is discussed in Chapter 4.

The name "hydrogen" given to the element is extremely befitting one, because it means a "water generator". Necessity of water in the formation of planet Earth was felt by Supernatural Power (the *Creator* of the *Universe*) who knew that plant and animal life will later dawn on the Earth. It sounds as though it was a pre-planned action. Whether life exists on the other planets is still not known with certainty. It is a matter of hot debate. Another important feature is the presence of a sphere of *ozone* around the Earth. This feature gives credence to the philosophy that most probably there is a *Creator* who had visualised that plant and animal life dawning later on the Earth will be required to be shielded from the harmful effects of the ultraviolet radiations that emanate from the Sun. The Earth possesses a magnetic core and a magnetic field around it. Dott and Batten (op. cit.) write, "the cause of the magnetic field has not yet been proved". Rotation of the Earth around the Sun and its

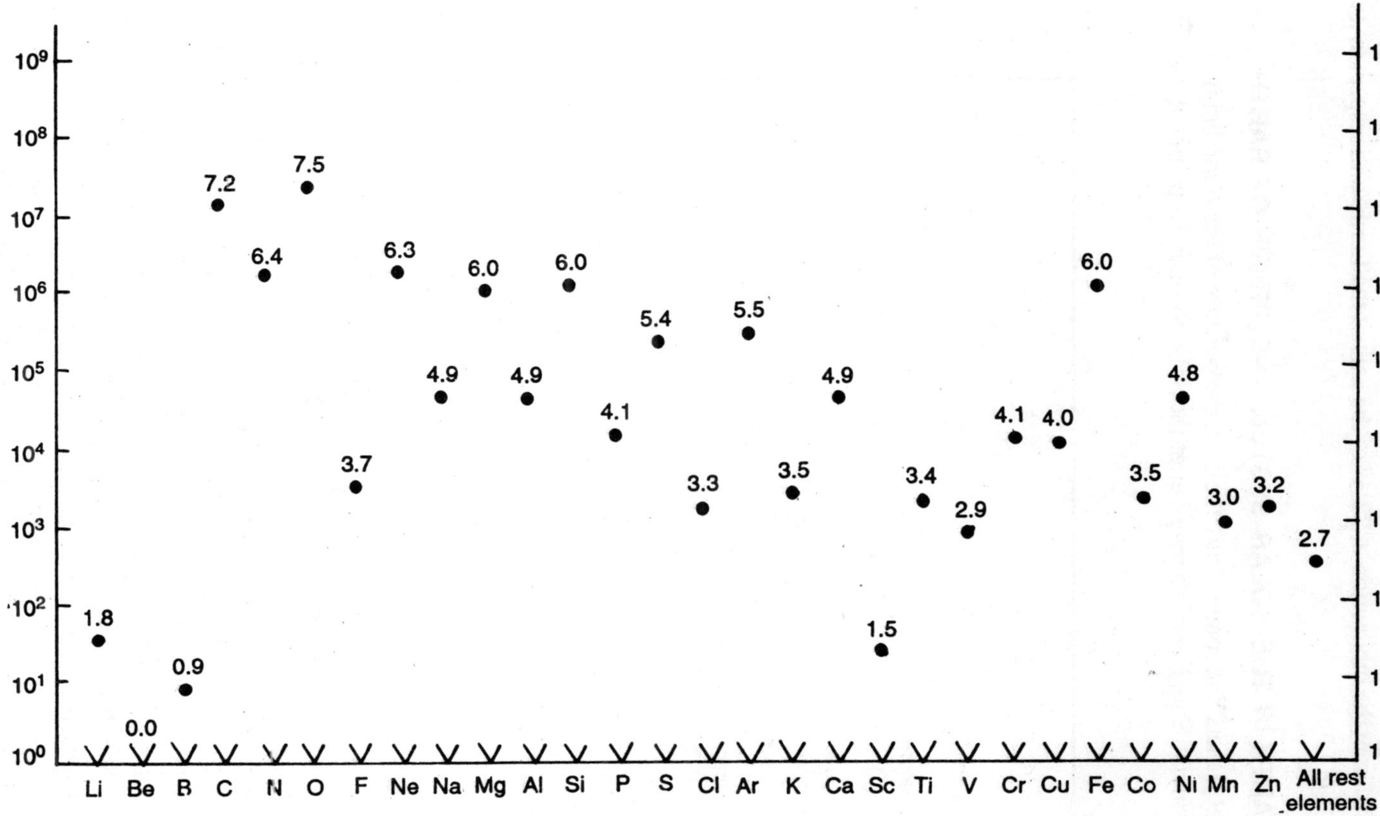

Fig. 3. Relative abundances (by weight) of the elements beyond Helium (After Fig. 19.9 of Krane, 1988).

revolution around itself, is the effect of the gravitational pull between them. The rotation helps to stabilize the position of the Earth in the space. However, it is not known as to which force brought about the inclination of 23½° of the axis of the Earth with respect to its orbital plane.

POSITION OF EARTH IN THE SOLAR SYSTEM AND ORIGIN OF EARTH

Earth is one of the 9 planets that rotate round the Sun, the position of the Earth being in the 3rd orbit (Fig. 4) from the Sun. Earth is a member of the Solar System: its parent, therefore, is the Sun.

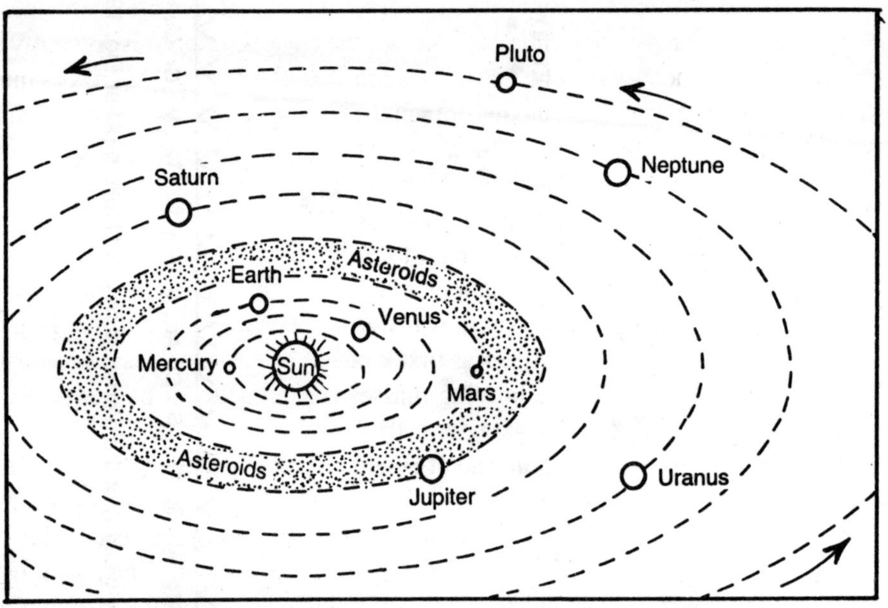

Fig. 4. Position of earth in the solar system (after figure from the Oxford School Atlas, 1978).

At present the Earth is in a solid state, at least externally, but its initial condition at birth, is visualised in two different forms. One view is that it sprang from the Sun in a gaseous state, which is described as the "Big Bang" hypothesis. With passage of time, the gaseous mass condensed to its present solid state. The second view is that there was a big sized meteorite in the space, which went on capturing big and small sized meteorites from the space, and grew bigger and bigger in size. Due to the increase in size, the core portion got compressed, which eventually led to the melting of the core. The process of melting then extended to the entire body, resulting in a molten mass. With the cessation of the act of capturing more meteorites from the space, the molten mass cooled to a solid body. This mechanism is described as the "Planetesimal hypothesis", its proponent being Chamberlain and Moulton. This hypothesis is based on the fact that, even today meteorites big and small, fall on the surface of the Earth. The dictum of "present is key to the past" is the logic behind the Planetesimal hypothesis. The lake Lonar, Buldhana, Maharashtra, India, is regarded to be a depression caused by the impact of a large sized meteorite on the Earth, which event had occurred in the geological past period. Carl Sagan (1985) writes, "The surface of the Moon is severely created by impacts from meteorites and these craters are not erased out, because Moon is airless and waterless. The Earth too must have been hit by meteorites creating numerous craters, but these have been wiped out by the geological agents like water, wind, ice and so on". Out of these two hypotheses, which actually led to the origin of the Earth, is still a topic of debate.

COSMIC PERSPECTIVE OF THE EARTH

The relation of the Earth to the cosmos is somewhat similar to that of the continents to the oceanic bodies. Continents have a shoreline. Unlike the oceans, cosmos is an empty space. Surface of the Earth is the shore of the *Cosmic Ocean*. As already noted, at the beginning of the Universe, there were no stars, galaxies or planets, but merely a uniform radiant fire-ball filling all the space. The chaos of the Big Bang was in a huge dense body made only of hydrogen (Big Bang means an explosion which took place due to reason unknown even today), resulted in several splinters, big and small, which were strewn into the empty space. The dense body of hydrogen was not burning prior to the point of Big Bang. For unknown ages after the explosive outpouring of matter and energy of the Big Bang, the cosmos was without form. Deep impenetrable darkness was everywhere. Here and there, globules of matter were condensing and nuclear fire was first kindled. Nuclear fuel was soon exhausted and the planets were formed. These burning masses and globules of hydrogen produced stars and several galaxies. One of the galaxies is the "milky way". In this milky way, Sun is one of the burning bodies. Sun is an inferno of H and He gas engaged in the thermonuclear reactions, flooding the Solar System with light. Big and small splinters thrown out from the Sun, were also burning, but soon the burning ceased and these gave rise to the several planets rotating around the Sun. Thus, the Solar System was formed. The positions taken by the several planets rotating round the Sun, are not dependent on the size, but are the outcome of the process of splitting away from the Sun.

CAUSE OF ELLIPTICAL ORBITS OF PLANETS AROUND THE SUN

According to the Kepler's law of planetary motion, a planet moves in an elliptical form, with the Sun at one of the foci, as shown in Fig. 5.

When the planet is nearest to the Sun, it speeds up. It slows down, when it is farthest from the Sun. Speed of the planets depends on their distance from the Sun. According to the Kepler's law, squares of the periods of the planets (period is the time taken to complete one orbit round the Sun) are proportional to the cubes of their distances from the Sun. More the distance, more slowly it rotates. The precise mathematical equation is given below :

$$P^2 = a^3$$

where, P is the period of rotation round the Sun, and a is the distance between the planet and the Sun.

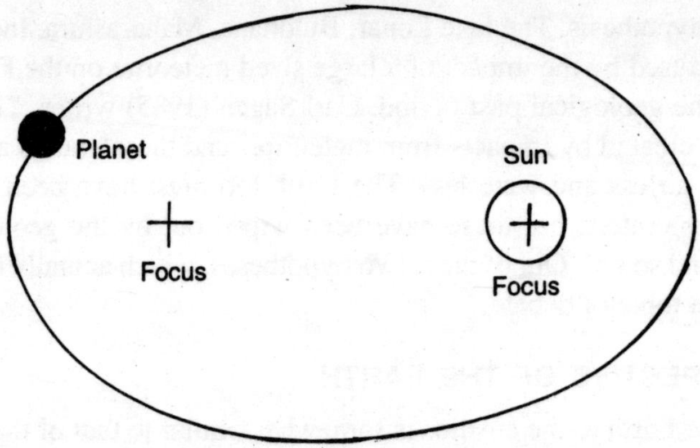

Fig. 5. Elliptical orbit of planets round the Sun with the Sun at one of its foci (after Ferris Timothy, "Coming of age in the milky way", 1990, affiliated East-West Press Pvt. Ltd., New Delhi).

For example, Jupiter is 5 units (one unit is equal to 15 crore km) away from the Sun. Substituting this value in the above equation, the period obtained is 11 years to complete one rotation round the Sun.

$P^2 = 5 \times 5 \times 5$, i.e., 125, and 125 is approximately the square of 11

The force of gravity declines as the square of the distance between two bodies. If the force of gravity was direct, and increased with distance, then the strongest force works on the most distant objects (planets), and the matter in the Universe would collapse into a single osmic lump. Gravity decreases with distance. Therefore, the planets move slowly when farther away from the Sun, and the speed increases when these are closer to the Sun. Due to this, the orbits of the planets around the Sun become elliptical in shape.

Newton's law of gravity predicts the orbit of the planets. The word "atom" in Greek means "indivisible", and "Proton" means "First". Every particle feels the force of gravity according to its mass or energy. Gravity is the weakest force. In the gravitational field, the force between two matter particles is pictured as being carried by a particle of "Spin 2" called "gravitron". This has no mass of its own. Therefore, the force that it carries is of long range. The gravitational force between the Sun and the Earth is ascribed to the exchange of gravitrons between the particles that make up these two bodies. This makes the Earth's orbit elliptical round the Sun.

CHAPTER

3

SOME FEATURES OF THE EARTH

From the cosmic phase of the Earth, let us descend to the planet Earth, and learn about its surficial and the internal features. Earth is placed in the third orbit (Fig. 4), a disposition that has made life sustainable on it. Is this disposition through a chance element, or there was some divine force responsible for it? It appears that the Earth conveys its reverence to the Sun through the act of rotation around it, like a devotee going around the statue of God in a temple. It is also significant to note that the Moon rotates around the Earth, which possibly indicates that the Moon sprang from the Earth, and is, therefore, rotating around it, like the Earth rotating round the Sun. These features clearly demonstrate that the Earth and the Moon have originated from the Sun.

The act of revolution of the Earth around its N-S axis, and rotation around the Sun, is due to the existence of gravitational force between it and the Sun. In the absence of such a force, the Earth and the Sun, both would have drifted away from one another, and lost in the enormous empty space of the Universe. The path of rotation around the Sun is elliptical (Figs. 4 and 5). An imaginary line joining the north and the south poles, which passes through the centre of the Earth, is called the axis. The locus of the path of rotation of the Earth round the Sun is called the orbit, which is elliptical. As a result, the Earth is twice nearer to, and twice farther away from the Sun, during one complete rotation round the Sun. These features are shown in Fig. 6.

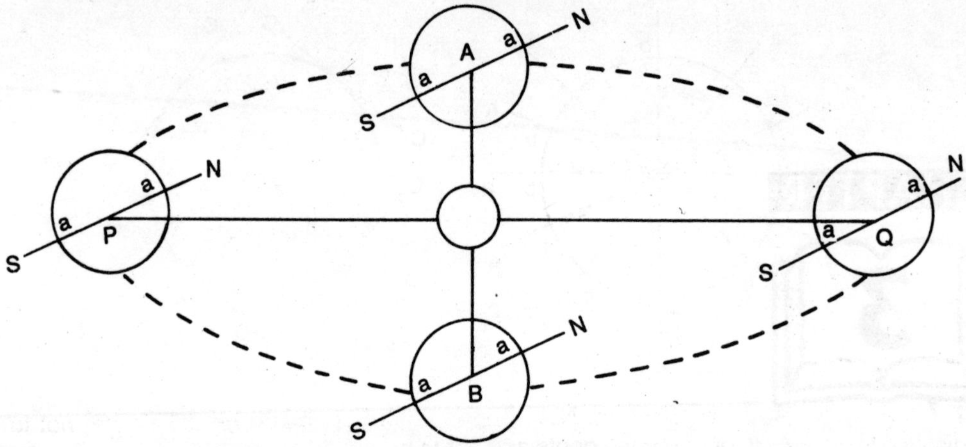

Fig. 6. Disposition of Earth during one complete rotation round the Sun. AQBP—Elliptical orbit of the Earth round the Sun. a-a : Axis of the Earth; A and B Closest positions of Earth from the Sun; A-B : Shorter axis of the orbit; P-Q : Longer axis of the orbit; P and Q : Farthest positions of Earth from the Sun.

According to the laws of planetary motion, velocity increases when bodies come closer, and decreases when they are farther from one another. Extending this law to the Earth and the Sun, it may be said that the Earth comes twice closer to the Sun, because of increase in its velocity. Likewise, it is twice farther away from the Sun, because its velocity decreases. It is also a fact that the gravity decreases with distance, and this brings about decrease in the velocity of the planets. Question that arises is which event is FIRST: Change in distance, or change in gravity? Does the Earth move closer to the Sun because of increase in gravity or the distance gets reduced, and therefore, the gravity increases? Whatever the cause might be, the path of rotation of the Earth round the Sun is rendered elliptical.

DISPOSITION OF AXIS OF EARTH IN THE ELLIPTICAL ORBIT

The attitude of the axis of the Earth needs discussion. It is often said that it is not vertical, but is inclined at angle of 23½°. However, this description becomes vague, unless the "plane of reference is indicated. The terms "vertical and horizontal" are meaningless for a body like the Earth, which is spherical, and is suspended in space. Even the surface of the oceanic body is not horizontal, but in totality it turns out to be curved one. In order to appreciate this feature, verticals and horizontals are drawn all over the surface of the Earth, and are presented in Figs. 7A and 7B.

Verticals when extended at depth, these meet at the centre of the Earth (lines aa in Fig. 7a). The verticals thus turn out to be radii of the Earth. Likewise, the horizontal lines (lines cc in Fig. 7b) become tangents to the circumference, which forms the surface of the Earth. Therefore, the verticals and the horizontals should be deemed as the *Radials* and the *Tangents*, respectively.

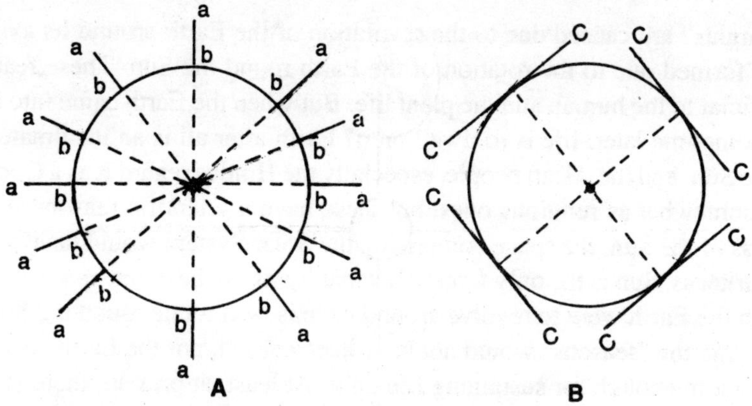

Fig. 7. (A) Verticals, (B) Horizontals. Note that lines a-b turn out to be radials and not to be verticals. Lines c-c turn out to be tangents and not to be horizontals.

The disposition of the axis of the Earth should be described with reference to the orbital plane formed by the rotation of the Earth around the Sun. The axis is not only contained in the orbital plane, but it is further inclined at an angle of 23½° with respect to the longer diameter of the elliptical plane. These features have been shown in Fig. 8.

A thought is likely to arise in mind regarding the relevance of dealing with the disposition of the axis of the Earth vis-a-vis the "Remote Control" in the mystique features of it. The

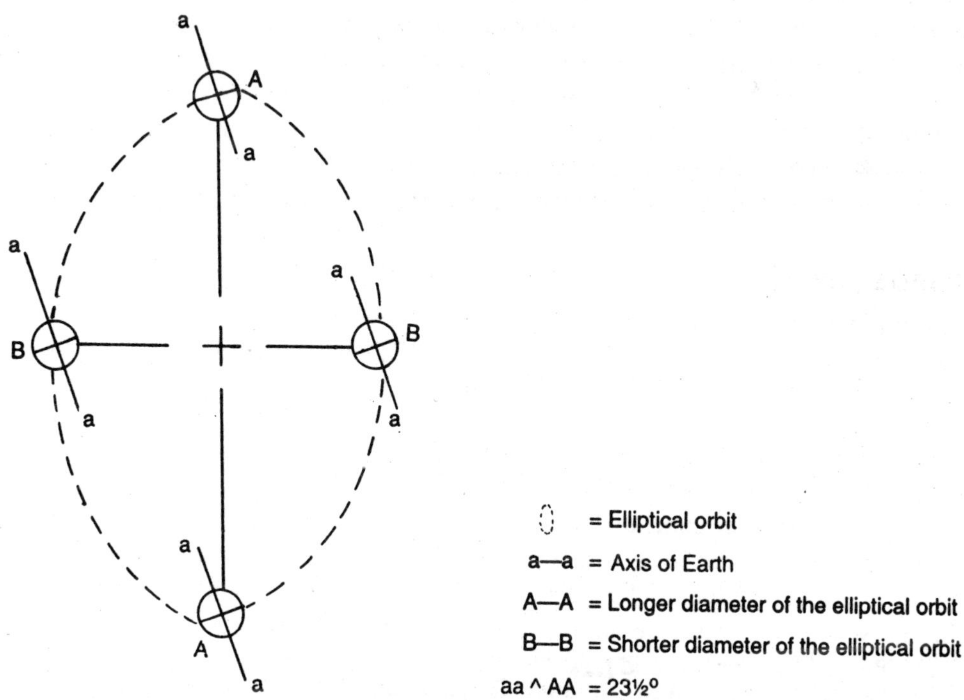

Fig. 8. Disposition of the axis of the Earth in its orbital elliptical plane.

"days and nights" are caused due to the revolution of the Earth around its axis, while the seasons are formed due to the rotation of the Earth round the Sun. These features are no doubt beneficial to the human and the plant life. But when the Earth came into being, did it know that sometime later, life is to dwell on it? Earth after all is an inanimate body. Man worships the Sun, and the Asian people, especially the Hindus regard it as a God. These two acts sound somewhat as religious ones, but these have a scientific reasoning. But for the burning mass of the Sun, the space comprising the Solar System would have been grouped in eternal darkness. Sun is the only *Lamp Post* that lights up the entire space surrounding it. Even though the Earth were to revolve around its axis, and rotate round the Sun, the "days and nights", and the "seasons" would not have been caused, nor the Earth could have been maintained warm enough for sustaining life on it. At least, at present, there is life only on the Earth. Does this not indicate that there is a *Remote Controller* (*Creator*) behind all these features of the Earth?

The role of a *Creator* is further evidenced by the fact that the Earth alone is having an envelope of ozone around it, that protects the plant and the animal life from the dreadful radiations emanating from the Sun, to which a mention has been made in Chapter 2. Earth possesses magnetic property. There is no specific shape to the magnetic body at the centre. It could be spherical, ellipsoidal or rod-shaped. It is held that the iron core spins and acts as a dynamo which produces electricity. What would be the adverse effect, if the Earth were not to possess the magnetic property? The intensity of magnetism is maximum at the north and the south poles. Why should the poles be developed at all? After all, any magnetic body attracts only iron particles. Does the magnetic property and the existence of poles, has any bearing on the stability of the Earth in the Solar System, as well as in the Universe? How the Earth acquired its magnetic core is not known with certainty. The axis of the magnetic N-S need not coincide with that of the geographic N-S. Why did the axis of the magnetic N-S change in the geological past, remains a matter of investigation. At least, at present, these features have remained unexplained, and therefore, these belong to the domain of mysticism.

According to Dott and Batten (op. cit.), the various elements encountered in the Earth were formed long before the formation of the protoplanet. Helium, though only second to hydrogen in the Universe, it is almost totally missing from the Earth, by loss to space. But is there hydrogen, which is lighter than helium, or that too has escaped into the space? This is because it was trapped by combining with oxygen to form heavy molecules (water). As noted in Chapter 2, all the elements are formed during the stage of burning of hydrogen through helium. Earth too must have gone through a similar process. The elements are initially in the gaseous state, and only on condensation, these combine to form "Oxides, Silicates, Sulphides" and so on. Water too must have been formed in this manner. Congealing and warming, the Earth released methane, ammonia and hydrogen gases that had been trapped within. The primitive atmosphere and the first oceans were formed. The global temperature of the Earth would have been below the freezing point of water, if not for the greenhouse effect. This effect keeps the oceans in the liquid state, and prevents the Earth from getting frozen, and creates congenial atmosphere for the living beings on the Earth.

The plant life was thoughtfully initiated by some Super Power (*Creator*) in order to save the animal life. The Earth has 90 atmospheres of CO_2, but it resides in the crust as limestones and other carbonates, and not in the atmosphere. If the Earth were to move a little towards the Sun, the temperature would increase slightly. This drives some of the CO_2 from the surface rocks, and generates stronger greenhouse effect. This in turn incremently heats the surface further. A hotter surface vaporises still more carbonates into CO_2, which causes run away greenhouse effect, resulting in very high temperatures.

Carl Sagan (1985) writes, "In the early days, lightning and ultraviolet light from the Sun, broke apart the simple hydrogen-rich molecules of the primitive atmosphere, the fragments spontaneously recombining into more and more complex molecules. One day, quite accidentally, a molecule arose. It began in the oceanic water. This molecule is the ancestor DNA (Deoxyribonucleic acid), the master molecule of life on Earth. The oceans were filled with simple green plants. Oxygen became a major constituent of the Earth's atmosphere, from its original hydrogen-rich character, and the Earth later acquired its oxidising atmosphere. Nitrogen is chemically inert. 99 per cent of the Earth's atmosphere is of biological origin. Life originated some 4.5 b.y. ago. Organic (life) molecules are essentially identical in all the plants and animals"

AGE OF THE EARTH AND THAT OF THE OTHER EPISODES

According to Condie (1984), "The earliest terrestrial crust appears to have formed between 4.2 and 4.5 b.y. ago, and the earliest continental crust between 3.8 and 4.0 b.y. ago. If the Earth had a primitive reducing atmosphere, it must have been lost in response to some catastrophic event in the first 200 m.y. after accretion. Rapid degassing of the Earth during the Archaean, produced an atmosphere rich in CO_2, N and H_2O. Oxygen is liberated into the atmosphere beginning at about 2.0 b.y. ago, and grows rapidly after 600 m.y., derived chiefly from photosynthesizing organisms. Composition of sea water has not changed significantly in the last 3.8 b.y., and that the volume of the oceans at 2.5 b.y. was about 90 per cent of the present volume. Warm climate existed in the pre-Cambrian and late mesozoic periods. Five worldwide glaciations have occurred at about 2300, 600-800, 500 and 300 m.y. ago, and during the Pleistocene. Life formed 3.8 b.y. ago in acqueous ponds on the Earth's surface by reactions of carbon-bearing compounds, in an oxygen-free environment, perhaps activated by ultraviolet radiations. Earliest fossil remains are organic microstructures and stromatolytes, which date back to about 3.5 b.y. Evolution of vertebrates (450 m.y.), vascular plants (425 m.y.), mammals (225 m.y.), flowering plants (150 m.y.) and man (3 m.y.)

EARTH'S MAGNETIC PROPERTY

(Given by T.N. Shivanand, *The Hindu, India's National Newspaper*, under the section of *Question Corner*, in the issue of 27.8.2003. Only relevant matter is given below).

The North Pole is near Ellef Ringes island in Northern Canada, about 1400 km South-

West of the North Pole. The South Pole is off the Coast of Ilkes Land, a part of Antarctica, about 2750 km South-East of South Pole. The Earth's magnetic field extends into space beyond the atmosphere, called magnetosphere. The belts surrounding the Earth are like doughnuts. The inner belt extends from about 1000 to 1500 km into the space. The outer belt reaches about 15,000 to 25,000 km into the space. The Earth's magnetic field traps particles and directs them towards the magnetic poles.

OPINION ABOUT MAGNETIC FIELD OF THE EARTH AT THE BEGINNING OF 21ST CENTURY

(Views of Prof. Rama Murthy of University of Minnesota, Geology and Geophysics on the magnetic field of the Earth have been given by *The Hindu, India's National Newspaper*, under the Section of Science and Technology, in the issue of 15.5.2003. Only relevant part is given below).

EVIDENCE FOR POTASSIUM AS MISSING HEAT SOURCE IN PLANETARY CORES

"There is a small problem with Earth's magnetic field; it should not have existed, as Earth's rock record indicates, it has, for the past 3.5 billion years. Motions in the Earth's molten iron core generate convection currents; which produce the field. Prof. Rama Murthy theorised that if radioactive elements such as potassium (a radioactive isotope of the element known as potassium-40), and perhaps uranium, thorium also exist in the core, the heat from their radioactivity could keep the core hot enough to move and maintain the magnetic field. A hot core will dissipate its heat one way or another. If it's only a little hot, the heat will be conducted in a smooth manner into the surrounding mantle. But if it's too hot, the heat energy drives convection currents in the core, turning it into a gigantic magnetic generator. Earth's magnetic field shields from the harmful cosmic rays and the constant stream of charged particles from the Sun known as the Solar Wind". Some scientists, however, doubt whether potassium could be present at all in the core of the Earth, because potassium is not found in any metal ores at the surface. It is also not possible to get samples from the core, to strengthen the views of Prof. Rama Murthy. Scientific probe thus has a limitation. How the Earth has acquired its magnetic core, continues to remain a mystery.

CHAPTER 4

EXTERNAL AND INTERNAL FEATURES OF THE EARTH

In the previous chapter, the Earth has been described as one of the planets of the Solar System. Let us now descend closer to the Earth, and understand its external and internal features. In order to acquaint ourselves with its shape and the surficial features, its dimensions have been set out in Table 1.

Table 1. Dimensions of Earth (data from Holmes, 1960)

Diameter	Equatorial	12,757 km (7,926.7 miles)
	Polar	12,714 km (7,900.0 miles)
Area	Sea Floor	361 million sq km (139.4 million sq. miles, 70.78%)
	The Land	149 million sq km (57.5 million sq. miles, 29.22%)
Greatest known height (above sea level) (Mt. Everest, India)		8840 meters (29,140 feet)
Average height (above sea level)		825 meters (2,707 feet)
Greatest known depth (below sea level) (Swire Deep)		10,800 meters (34,430 feet)
Average depth (below sea level)		3,500 meters (12,460 feet)
Volume		1,02,000 million cubic km (259,100 million cubic miles)

From the data set out in Table 1, it is seen that the Earth is not a spherical body. Photographs taken from the satellites, give an impression of it to be spherical in shape, and the surface to be smooth. The Solar and the Lunar eclipses also create an impression that the Earth is spherical in shape, from the circular shape of the shadow produced by the eclipse. However, the dimensions of the Earth do not support such a shape. The difference between the polar and the equitorial diameters is 43 km. (26.7 miles) indicating thereby that the Earth is somewhat pressed at the two polar regions. The ruggedness of the surface of the Earth is concealed by the extensive stretch of the Pacific ocean in the main, and assisted by the Atlantic and the Indian oceans. The values of the greatest height and depth attained by the surface of the Earth can be appreciated quickly by removing the cover of the oceanic water. The difference or the separation between the greatest height (8840 meters, or 29,140 feet) and depth (10,800 meters or 34,430 feet), is as much as 19,640 meters (8840 + 10, 800), or 63,570 feet (29,140 + 34,430). Even if the average values be considered, the difference is of 4325 meters (825 + 3500). These figures clearly indicate the ruggedness of the surface, and a non-spherical shape of the Earth. In order to appreciate the ruggedness of the surface, and the true shape of the Earth, Figs. 9A, B, C, D have been drawn. Fig. 9D is the

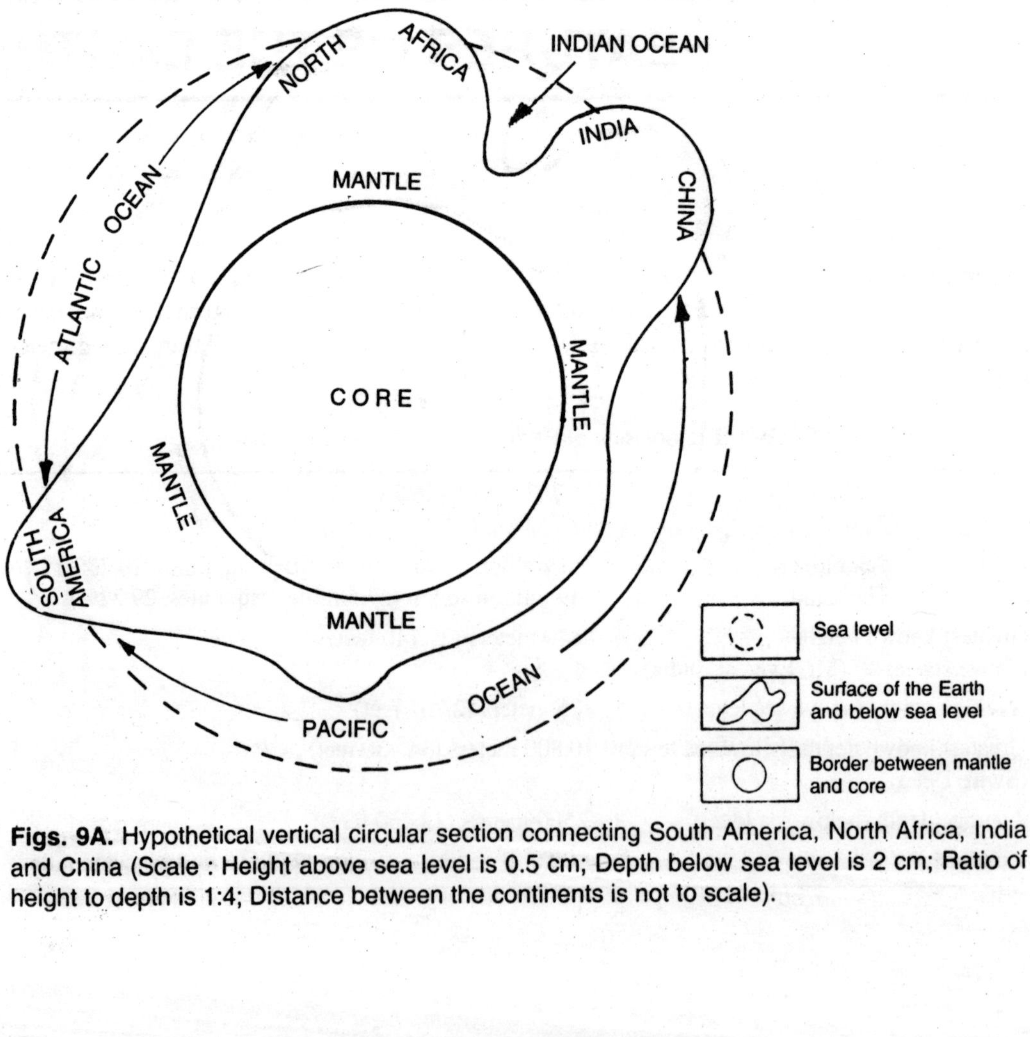

Figs. 9A. Hypothetical vertical circular section connecting South America, North Africa, India and China (Scale : Height above sea level is 0.5 cm; Depth below sea level is 2 cm; Ratio of height to depth is 1:4; Distance between the continents is not to scale).

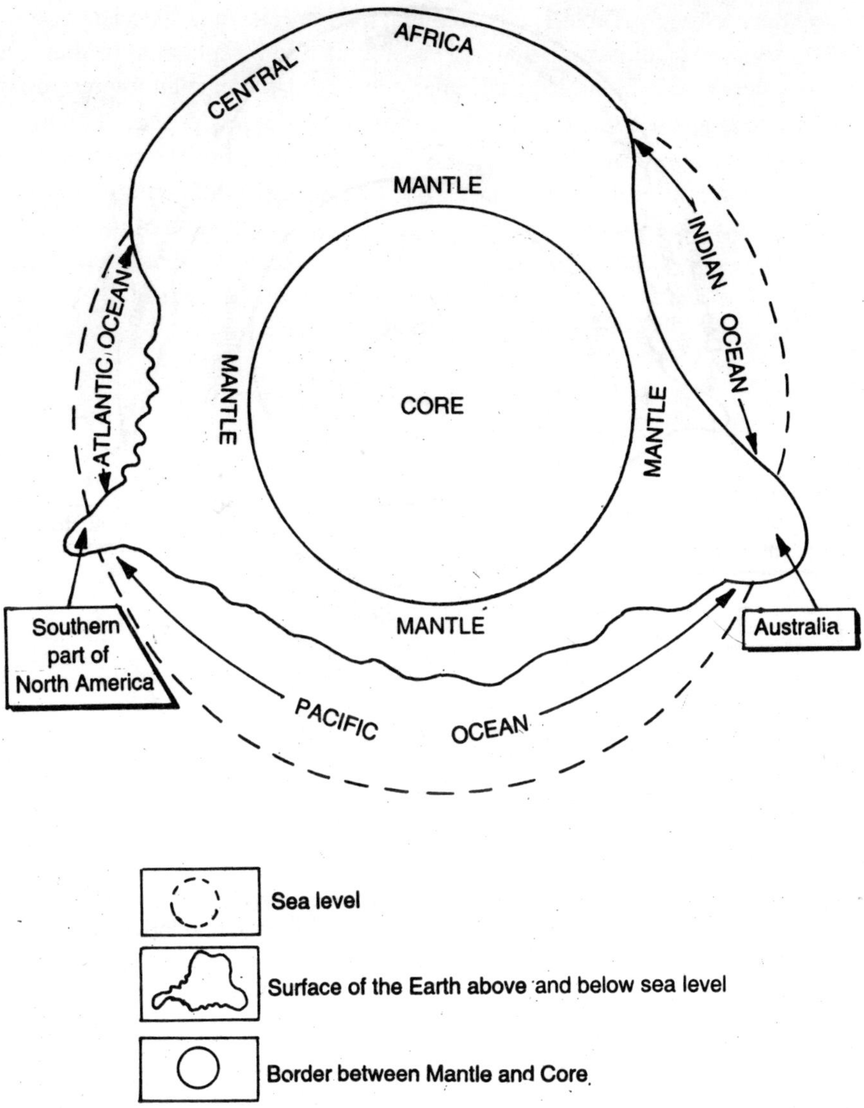

Fig. 9B. Hypothetical vertical circular section connecting southern part of North America, Central Africa and Australia (Scale : Height above sea level is 0.5 cm; Depth below sea level is 2 cm; Ratio of height to depth is 1:4; Distance between the continents is not to scale).

actual profile of sea floor between Newport Rhode island and Gibralter. This figure decidedly conveys that the sea floor is extremely rugged, and there is great variation in heights and depths.

The shape of the Earth is sometimes described to be similar to that of an orange, because that fruit is depressed at its opposite two ends, and its surface is wrinkled. However, instead of comparing the shape with some other object, it is better to describe it as "Earth

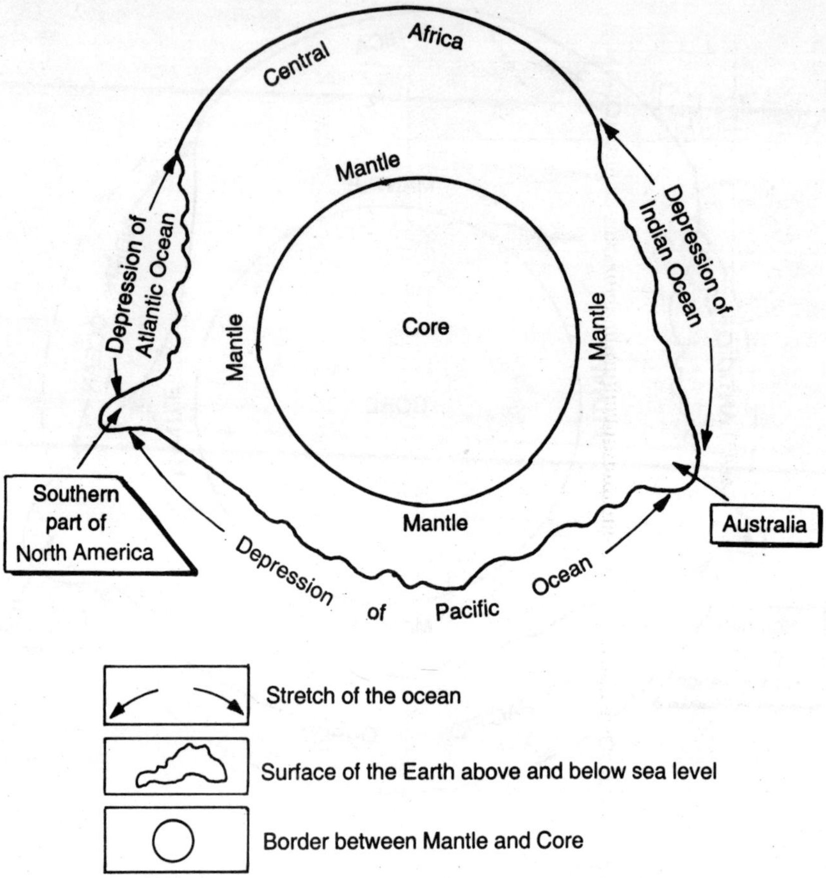

Fig. 9C. Hypothetical vertical circular section connecting southern part of North America, Central Africa and Australia, on removing the cover of Atlantic, Pacific and Indian oceans. Note that the appearance of the surface of the Earth is very rugged and far from spherical one in shape (Scale : Height above sea level is 0.5 cm; Depth below sea level is 2 cm; Ratio of height to depth is 1:4; distance between the continents is not to scale).

shaped". Even then this feature of the Earth needs explanation pertaining to its depression at the two polar regions. Is it the outcome of the spinning around the N-S axis, or has it any other significance?

INTERNAL FEATURES OF THE EARTH

Let us now turn to the internal features of the Earth. Its composition from the surface to the core has been described by several scientists in different forms. Some of them have been set out in Figs. 10A, B, C, D.

The thickness of the crust and the other parameters of the Earth noted by Richter (1969), Schlee (1975) and Condie (1984) are set out in Table 2.

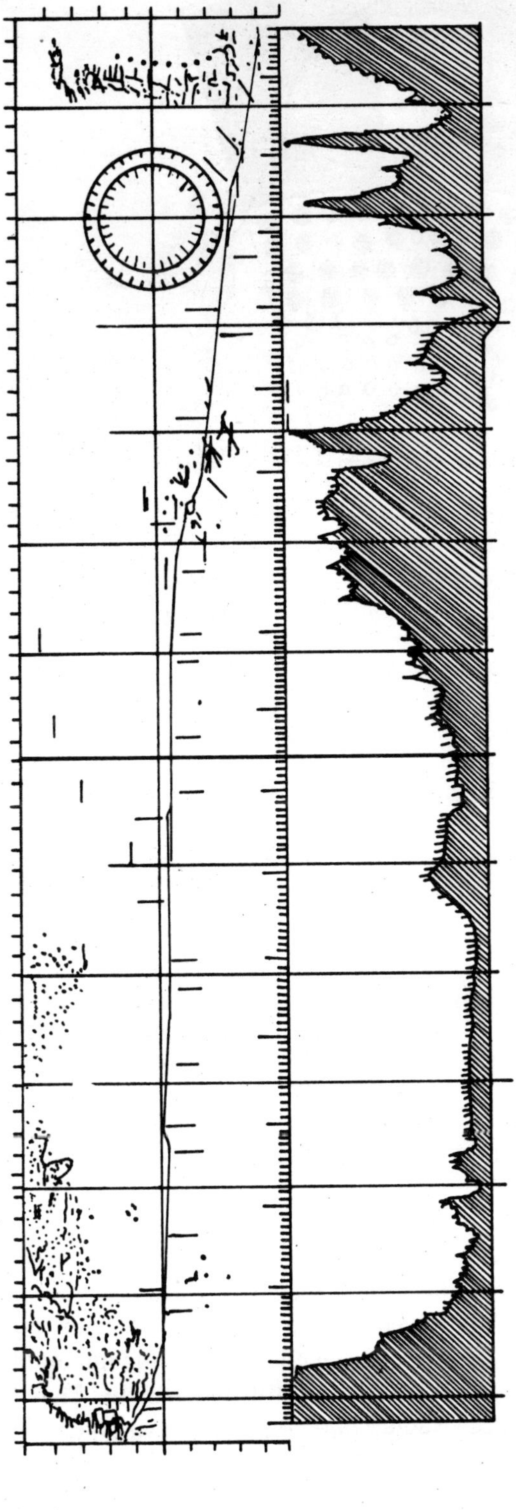

Fig. 9D. Actual continuous profile of seaf loor from Newport, Rhode Island, to Gibraltar (after Schlee, 1975, figure on page 251).

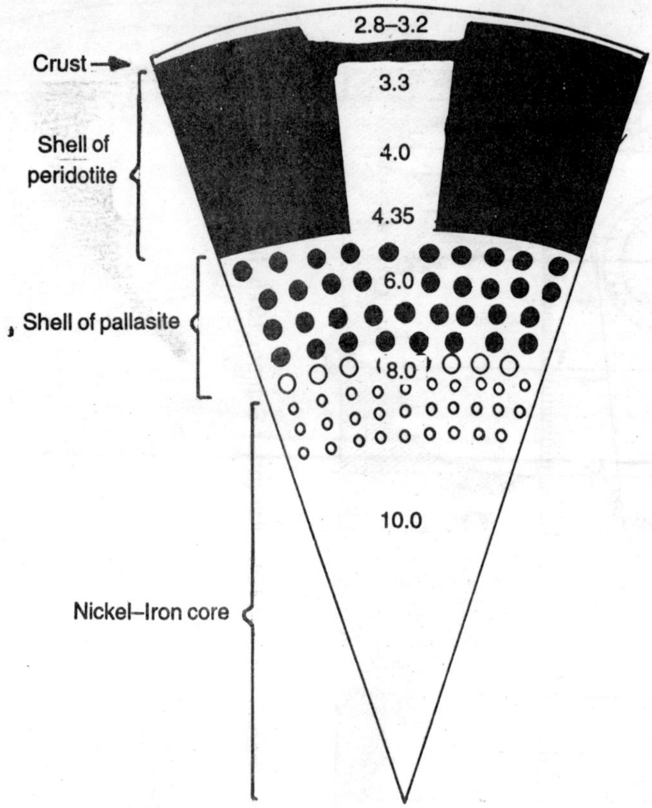

Fig. 10A. Section through the Earth (According to Williamson and Adams) (After Fig. 1A of Tyrrell, G.W., 1929, p. 9).

Layer	Rocks	Thickness	Density
Crust	Granite and basalt	100.0 km	2.8-3.2
Shell of peridotite	Peridotite	1500.0 km	3.3-4.35
Shell of pallasite	Siderolite plus increasing admixture of Ni-Fe	1400.0 km	6.0-8.0
Nickel-Iron core	Partly metallic Ni-Fe	3400.0 km	10.0
Radius of Earth	—	6500.0 km	—

(Analogy of smelting of copper ores is extended).

Note that no sharp boundaries are shown by Williamson and Adams unlike that done by Goldschmidt. Rocks become denser gradually and not sharply.

Features presented in Figs. 9A, B, C, D and 10A, B, C, D and the data set out in Table 2 indicate that there is considerable variation in the values of density, thickness of layers, models of the internal constitution of the Earth, as given by scientists. The nomenclature of the layers is also different from scientist to scientist, causing confusion in the mind of the readers, because the different names are in fact the synonyms of one another. Thus the terms "lithosphere, crust, sial" are synonymous. Though there is a marked variation in the

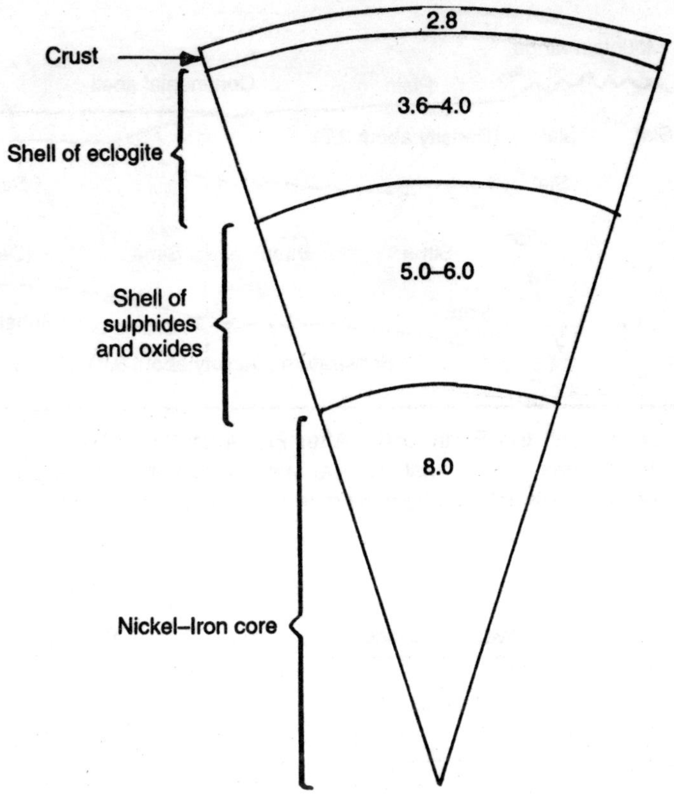

Figs. 10B. Section through the Earth (According to Goldschmidt) (After Fig. 1B of Tyrrell, G.W., 1929, p. 9)

Layer	Rocks	Thickness	Density
Crust	Granite and basalt	120.0 km	2.8
Shell of eclogite	Eclogite (dense silicates)	1000.0 km	3.6-4.0
Shell of sulphides and oxides	Sulphides and oxides	1700.0 km	5.0-6.0
Nickel-Iron core	Nickel-Iron	3400.0 km	8.0
Radius of Earth	—	6200.0 km	—

(Analogy of smelting of iron ores is extended.)

Note that the sharp boundaries and the rocks become denser sharply.

thickness of the layers, yet the Earth is projected to be composed of concentric spheres (layers) of uniform thickness. Likewise sharp lines of demarcation are drawn on the basis of density of rocks. However, the densities of the rocks vary gradually from 2.6 to 8.0 towards the centre of the Earth (Figs. 10A, B, C, D). Rocks composing the "lithosphere/crust/sial/sima" are igneous that make up 95 per cent, the remaining 5 per cent comprising the sedimentary and the metamorphic categories. The igneous rocks are not at all layered ones, and hence do not possess uniform thickness.

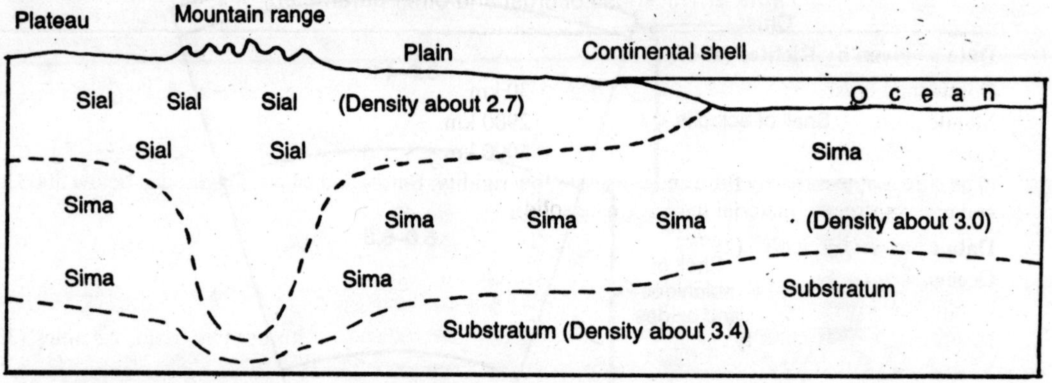

Fig. 10C. Section through the Earth crust (After Fig. 4 of Arthur Holmes, 1960). Note : The thickness of Sial and Sima is not constant. Sial layer is very thin below the continental slope, and it is absent beneath the oceanic floor, and its place is taken by sima.

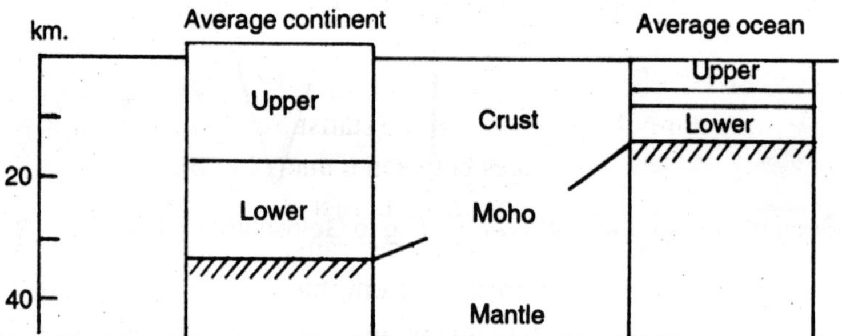

Fig. 10D. Average cross-sections through continental and ocean crust (After Fig. 13.2 of Park, R.G., 1989).

Details of Fig. 10D

Layer	Average thickness in km	Mean density
Continental crust		
Upper part	18.0	2.7
Lower part	15.0	2.9
Mohorovicic discontinuity below continental crust at a depth of 33 km		
Ocean crust		
Upper part	7.0	2.6
Lower part	8.0	2.9
Mohorovicic discontinuity below oceanic crust at a depth of 15 km		
Mantle		3.4

Note that the crustal layer is not uniform in thickness. The continental crust is nearly two times thicker than the oceanic crust.

Table 2. Thickness of crust and other parameters of Earth

Data as given by Richter (1969)	
Continental crust	30 km
Mantle crust	2900 km
Core	5000 km
(The core is at least partly fluid and possesses low rigidity. Bullen and others suggest that below 5000 km, and upto centre, the material may again be solid).	
Data as given by Schlee (1975)	
Oceanic crust	4 miles (6.0 km)
Continental crust	22 miles (33.0 km)
Mohorovicic discontinuity	4 miles (6.0 km) beneath oceanic basin, 22 miles (33.0 km) beneath continents
Data as given by Condie (1984)	
Lithosphere	50-150 km (upper 6-40 km designated as crust)
Mohorovicic discontinuity	10-12 km beneath oceans, 30-35 km beneath continents
Mantle	2900 km
Core	5200 km
Radius of the Earth	8250 km
(Lithosphere rests on hotter, denser layer called Asthenosphere. It yields plastically).	

Most salient features emerging out of the statistics set out in Figs. 10A, B, C, D are that:

(i) thickness of the crust ranges between 6 and 120 km;

(ii) average thickness of the continental crust ranges between 30 and 35 km;

(iii) average thickness of the oceanic crust ranges between 6 and 12 km;

(iv) 60 per cent of the entire crust is 15 km thick and is occupied by the oceans, while 40 per cent of the entire crust is 80 km thick constituting the continents.

It is thus quite apparent that the crust of the Earth is not uniformly thick, and is not made up of layers, as is generally conceived and projected. In Figs. 11A, B, the traditional and the more realistic projection of the constitution of the Earth has been presented.

Let us now consider the parameter of the density of the rocks. In Table 3, the values of density alongside with the rocks occurring in the crust, the mantle and the core are presented.

Table 3. Density of crust, mantle and core

Part of the Earth		*Name of the Rock*	*Density*
Crust		Granite, basalt, sedimentary rocks, Metamorphic rocks	2.6-2.9
Mantle		Peridotite, eclogite, pallasite, sulphides	3.4
Core	Outer part	Liquid state and made up of Fe-Ni, contains 5-15 per cent of one or more "low atomic weight elements"	8.0
	Inner part	Alloy of Fe-Ni, practically molten or near melting temperature	
Entire Earth			5.0-5.27

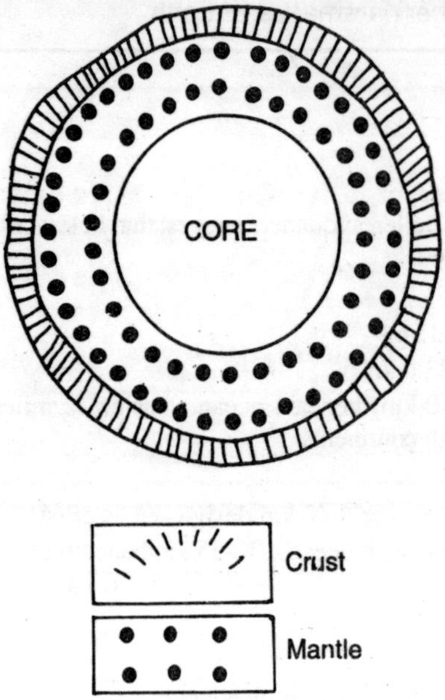

Fig. 11A. Traditional projection of constitution of the Earth. Note that the thickness of the crust and the mantle is uniform.

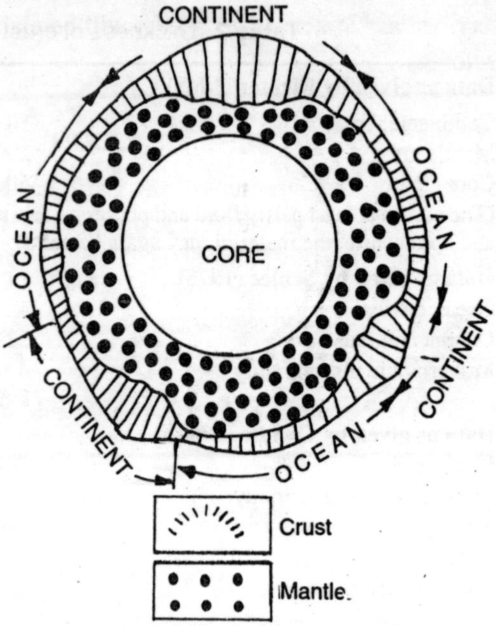

Fig. 11B. Realistic projection of constitution of the Earth. Note that the unequal thickness of the crust affects that of the mantle.

The density of innumerable samples of rocks collected from several parts of the crust, and from the mines, has been determined. The density of the material composing the core and the mantle, however, is deducted through logic. The density of the Earth is estimated to be 5.527, while that of the crust ranges between 2.6 and 3.4. Therefore, the core should be necessarily made up of materials denser than that of the crustal rocks, in order to attain the overall value of density of 5.527 of the entire Earth. Iron and nickle are, therefore, inferred to compose the *Core* as these are heavy elements. However, noble metal Gold has not been projected to constitute the *Core*, though it is the heaviest metal. Another feature of importance is that the density of the material (rocks) steadily increases from the crust through the mantle towards the central part of the Earth. Therefore, sharp breaks between the *Crust* and the *Mantle*, and that between the *Mantle* and the *Core*, are unnatural, because the Earth has originated through the process of crystallisation of its primitive molten state. Such a process produces rocks having imperceptible boundaries between one another. As a consequence, the crustal rocks will grade imperceptibly into those of the mantle, and those of the mantle with materials of the core. Such a projection of the internal structure of the Earth is more plausible because it is logical and realistic one. Out of the several types of internal constitutions shown in Figs. 10A, B, C, D, that of Williamson and Adams (Fig. 10A) is realistic, as they have projected a gradational and not a sharp change in the density of the rocks towards the core of the Earth. The non-uniform thickness of the crustal rocks (layers) will have an impact on the shape and the size of the continents, that are often described as "Plates". This aspect will be elaborated while dealing with the topic of "Plate Tectonics".

INTERNAL STRUCTURE OF MOON

So far the internal structure of the Earth has been described. It is felt appropriate to generally deal with that of the Moon, because the Moon has a genetical relation with the Earth. The composition of the Moon as described by Condie (op. cit.) is given in Table 4.

Table 4. Composition of Moon (after Condie, 1984)

Layer	*Thickness in km*	*Composition*	*Age*
Crust	60-400	Gabbroic anorthosite	
Upper mantle	300-400	Pyroxenes and olivines	4.46 b.y.
Lower mantle	500	Olivine dominating plus pyroxenes	
Partly melted mantle	400 ± 100		
Core	200-400		
Radius of Moon	1800		

A comparison of the data set out in Table 4 with that of the Earth indicates that the Moon is considerably small, it being about 1/3 in size. The glaring difference is that its crust is devoid of acidic igneous rocks like the granites, the granodiorites, the syenites and so on. In other words it means that the crustal density of the Moon is around 3.4 as against 2.7 of the Earth. It also indicates that the Moon is densely packed than the Earth. One more significant feature is the age. It is believed that most of the lunar rocks were formed during a major magmatic event at about 4.46 b.y. ago as noted by Condie (op. cit., p. 18, Burwah, 1969, and Condie, 1981a). Thus the lunar crustal rocks are very much older than those of the Earth. Does that mean that the Moon is not an offspring of the Earth? The core of the Moon is not described to be made up of metals like that of the Earth. In all probability, the Moon may not be possessing a magnetic core.

CHEMICAL COMPOSITION OF THE EARTH

Having described the density and the thickness of the layers composing the Earth, let us turn our attention to the parameter of its chemical composition. This aspect belongs to the realm of petrology. The crust of the Earth is made up of the igneous, the sedimentary and the metamorphic rocks. The igneous rocks alone formulate 95%, the other two varieties together make up only 5% of the crust of the Earth. Further, the sedimentary and the metamorphic rocks are derived from the igneous rocks. Therefore, the chemical composition of the crust will be generally reflected by that of the igneous rocks. The average chemical composition of the crust upto 10 miles depth as given by Tyrrell (1929) is given in Table 5.

Table 5. Average chemical composition of igneous rocks (after Tyrrell, 1929, p. 47)

SiO_2	59.12
Al_2O_3	15.34
Fe_2O_3	3.08
FeO	3.80
MgO	3.49
CaO	5.08
Na_2O	3.84
K_2O	3.13
H_2O	1.15
TiO_2	1.05
P_2O_5	0.30
MnO	0.11
Rest	0.02
Total	100.00

The chemical composition of the rocks composing the mantle is more basic than that of the crustal rocks. But what is thought-provoking is the fact that the content of iron is about 6-8 per cent in the crust and the mantle, while the *Core* having a radius of 3400 km is regarded to be entirely made up of iron and nickel. Goldschmidt and Williamson have extended the analogy of the process of smelting of ores (copper, iron) to the formation of the Crust, Mantle and the Core. The Earth, however, has come into being from a molten silicate state (magma) through the process of crystallisation. The first formed high temperature minerals from the melt are olivine, pyroxene, which consume Fe, Mg and Ca. Thus, the rocks like the peridotite, picrite, dunite should have built up the core of the Earth, and not merely by Fe and Ni (iron and nickel).

At this stage it is felt relevant to consider the modus operandi of formation of the elements encountered in the Earth. Table 5 presents the chemical composition of the crustal rocks, the core being regarded to be made up of iron and nickel. It is, however, felt necessary to work out the chemical composition of the entire Earth, when it was in the molten state. Whereas the chemical composition of the crust has been determined by analysing numerous samples of rocks, such a procedure, however, cannot be extended to the deeper parts and especially to the core portion. The inference of iron and nickel (Fe, Ni) constituting the deeper parts and the core is based on :

(a) the magnetic property possessed by the Earth, and

(b) the necessity of projecting denser minerals (magnetite, hematite, ilmenite, etc.) to bridge up the difference between the density of the lighter crust (density 2.6-3.0) and the overall density of 5.527 of the entire Earth.

The planets acquire their chemical composition (elements) through the thermonuclear reactions operative during the burning of hydrogen and helium at very high temperature of 10-12 million degrees Kelvin. *The quality and the quantity of the elements do not change, once the gaseous mass is thrown out from the burning body of the stars* (*Sun*). It merely cools to a molten state, and then crystallises to form solid matter (rocks) of the planets. The Earth too must have gone through similar ordeal of events. Krane (1988) has set out the values of abundance of elements encountered in the stars, and those have been presented in Table 6.

Table 6. Order of abundance of elements in stars (derived from Fig. 3)

Value of abundance	*Elements*
$10^{7.5}$	O
$10^{7.2}$	C
$10^{6.4}$	N
$10^{6.0}$	Si, Mg, Fe
$10^{5.4}$	S
$10^{4.9}$	Al, Na, Ca, Ni
$10^{4.1}$	P
$10^{3.9}$	Cu
$10^{3.5}$	K, Co
$10^{3.4}$	Ti
$10^{3.0}$	Mn

Scrutiny of Table 6 reveals that oxygen is the most abundant element, the least being Mn. Oxygen, carbon and nitrogen though abundant, these occur in the gaseous state. Si (Silicon) in combination with C, Al, Na, Ca, K, Fe, Mg and Ti, builds up the silicate minerals of the crustal rocks. Heavy elements like Fe, Ni and Ti, build up the minerals of the deeper parts and the core of the Earth. Oxygen, nitrogen and carbon (in the form of CO_2) build up the atmosphere.

The volume of rocks formed from the melt will be proportional to the volume of the minerals. Thc volume of the minerals formed, in turn, will depend upon the abundance of the different elements. It is, therefore, logical and reasonable to expect the abundance of the elements occurring in the Earth when it was in the gaseous state, to be more or less similar to those described by Krane (op. cit.) for the stars, because there is no process by which the elements could be lost, except hydrogen and helium.

The fundamental element around which the minerals composing the crustal rocks are composed is silicon (Si), whose value of abundance is 10^6. Likewise the critical elements building up the minerals of the deeper parts and the core of the Earth is iron (Fe) whose value of abundance is also 10^6. With identical values of abundances, Si has, however, given rise to rocks (crust) having a thickness of 150 km, while the remainder thickness of 6500 km of the Earth, upto the core, is built up mainly by iron (Fe) and nickel (Ni). The ratio between the volume of material (rocks) produced by Si and Fe-Ni works out as 3:27. Such a glaring discrepancy of unequal thickness and volume resulting from identical values of abundance of elements, certainly deserves investigation.

Now turning our attention to calculating the chemical composition of the entire Earth when it was in the molten state, the dominance of SiO_2 in the crustal rocks, and the monopoly of iron in the core, are of importance. SiO_2 makes up 59.12 per cent (Table 5), while iron (Fe) may reasonably be presumed to constitute 90 per cent of the core, the remaining 10 per cent to be made up of Ni, Ti. It is logical to speculate the presence of Pt, Pb and Au, because these are heavier than Fe and Ni, and these will contribute to build up a denser core. It is also felt appropriate to rename the core as "*Core* of iron, nickel, platinum, gold and lead", rather than the popular designation as "Ni-Fe *Core*"

On mixing up the two dominant elements, SiO_2 of the crust and Fe of the core, the mixture will consist of (on rounding off 59.12 to 59) 59 parts of SiO_2, and 90 parts of iron (Fe), in a volume of 200. The percentage values of SiO_2 and of iron will work out as 29.5 and 45, respectively. The chemical composition of the entire Earth may arbitrarily be considered to be 45% iron (Fe), 29.5% SiO_2 and 25.5% remaining oxides (Al_2O_3, Na_2O, MgO, K_2O and so on)

The chemical composition of the Earth raises a question. What is the cause of production of more Si and Fe during the thermonuclear reactions? Further, Fe is 1½ times more than SiO_2. The formation of elements from hydrogen and helium belong to two categories: (a) the iron group elements are formed by the non-reversible endothermic reactions, (b) the rest of the elements are formed by the reversible exothermic reactions. Could this be the cause of super abundance of Fe, and dominance of Si in the Earth?

CHAPTER 5

SHAPE, SIZE, NUMBER AND DISPOSITION OF CONTINENTS

Continents are those portions of the Earth's crust that are above the sea level. The seas and the oceans are formed only because the surface of the Earth is rugged, resulting in the development of elevations and depressions. In order to visualise this aspect, Fig. 9C has been drawn which shows how the surface of the Earth would appear, if the seas and the oceans are withdrawn. The feature to be noted is that the continents apparently appear to be independent bodies. But these in fact are all connected to one another through the continental slopes and the floor of the seas and the oceans. The feature to be stressed is that the continents are not detached from one another, and the entire surface of the Earth is a continuous entity. Of the total surface (of the Earth) of 510 million sq.km., the land surface is a mere 149 million sq.km., while the oceanic-sea floor constitutes 361 million sq.km. In terms of percentage, the oceanic sea floor is 70.78, and the land surface is as little as 29.22. Thus all the continents taken together are spread over a very small area.

The area available for agricultural cultivation and for man's habitation gets reduced, because of forests, mountainous tracts, ice caps (the north and the south polar regions called as Arctic and Antarctic continents), the inland-lakes, the deserts, and so on). Therefore, man is trying to reclaim land by pushing the sea and the oceanic water backwards. There is acute shortage of space for human habitation owing to the "population explosion". Multi-storied housing complexes are springing up. If the creation of land and the oceans is to be deemed as a natural process, it would have been better, if the ratio between the land and the sea was 2.5:1, or at least 1:1, and not 1:25, as at present. Such a distribution would have eased the shortage of space for the human habitation, currently faced by the mankind! But the distribution of the land and the sea and oceans as at present, appears to be a thoughtful action. As water is essential to the animals, the plants, and the man, the extent of water (oceans and seas) is probably, therefore, more than that of the land. Is this feature to be attributed to the outcome of a natural phenomenon, or to some Supernatural Power - the "*Creator*" of the Universe? Whether the surface of the Earth was primordially (i.e., when the Earth was born rugged or the ruggedness was produced by the action of geological agents, like the rains, the rivers, the wind and so on, is a debatable issue. Uncertainty, therefore, exists in this regard. Judging from the temperature of formation of the minerals as given by Dott and Batten (op. cit.), minerals composing the different rocks, are formed within the temperature range of 300°C-1500°C, while water is formed only below 100° centigrade. Therefore, the surface of the Earth must have been built earlier than the formation of water, in the form of rains. The surface of the Earth must have been without the spread of the seas and the oceans, a situation which is unlike the one presently noticeable on the surface of the Earth.

Let us now turn our attention to the shape, the number, the size and the disposition of the continents. The continents could be formed only if the land-surface comprises of elevations and depressions, coupled with large volumes of water to occupy the depressions. Regarding the number of continents, uncertainty prevails. Did the Earth commence with all of the present-day continents, or the situation was different in the geological past? Perret (1976) writes (p. 93), "The first landmass was located around the south pole. It broke into two parts; Antarctica which remained at the same location and another mass which drifted and broke into several parts, the present continents, forming the various seas and oceans. The drifting is continuing and the end result will be the formation of new landmass located around the north pole". Condie (op. cit.) has opined that "two or more super continents existed by 1000 m.y. ago. During the late pre-Cambrian and Cambrian, Laurasia fragmented into 5 or more sub-continents. These in turn collided reforming Laurasia, which combined with Gondwana in the late Palaeozoic to form Pangaea".

Archaeological research has yielded clue to ascertain the disposition and the number of continents in the bygone years. A study of mankind's genetic family tree has come up with a distinctly hereditical conclusion - Adam and Eve may never have met. Researchers have traced the male sex chromosome found in all men alive today to a Stone Age ancestor who lived 50,000 years ago. The dating of this Adam - reported by Dr. Peter Underhill of Stanford University and colleagues in Nature Genetics - provides strong evidence for the theory that modern humans evolved in Africa and dispersed around the World about 50,000 years ago. "The fascinating thing is that this confirms that our species dispersed out of Africa which is what we already had a good inkling of, from earlier studies of mitochondrial DNA", said Prof. Colin Renfrew of the McDonald Institute for Archaeological Research in Cambridge (taken from report appearing in *The Hindu, India's National Newspaper*, dated 1st November, 2000).

Palaeontologists working in Kenya have found the fossilised remains of a human face and skull thought to be some 3.5 m.y. old which will shed new light on man's evolution, the journal *Nature* reported. Mr. Museums of Kenya and colleagues found a battered but almost complete skull and face of an entirely new breed of early human.

From the title "The Journey of Man" written by Dr. Spencer Wells, Telegraph Group Limited, London, *The Hindu, India's National Newspaper*, dated 21.12.2002, has reported interesting matter, relevant part of which is given here. "London 20.12.2002; All human DNA coalesces to a single ancestor living in Africa around 60,000 years ago. The original African closest living relatives are the San Bushmen who now live in Namibia, Botswana and the northern part of South Africa. Sometime around 50,000-60,000 B.C., a group of Africans started walking. In those days, sea levels were far lower than they are now and there was virtually unbroken coastline all the way from Africa to Australia. The colonising humans divided; up through Egypt into West Asia and on to Central Asia; west across the Steppes into Europe; into China and to Siberia".

Can these findings of Dr. Peter Underhill, Prof. Colin Renfrew, Mr. Museums, and of Dr. Spencer Wells be taken as the latest evidence to infer that there was a singular, giant landmass (continent), Africa being its centre? The concept of Pangaea thus gains new evidence. Man could not have migrated to different parts of the world, if oceans and seas had separated them. It is also possible that there might have existed land bridges between the distantly located landmasses (continents), enabling the man to migrate to different continents. But this will create a new problem. What happened to the land bridges, because those are not to be seen today. Their disappearance requires scientific explanation.

NUMBER OF CONTINENTS

Landmasses rising above the sea level assume the status of continents, of course their sizes are also required to be taken into consideration. Therefore, the number will depend both upon the sea level, as well as on the elevation attained by the landmasses. Lower the sea level, greater will be the number of the continents, higher the sea level, lesser will be the number of continents. The number at present is 7, the names being, South America, North America, Greenland, Europe, Africa, Asia, and Australia. The Arctic and the Antarctic landmasses being permanently capped by thick sheet of ice, are not designated as continents. However, what was the number of the continents in the geological past? During the pre-Cambrian period, there were only two continents; the Gondwanaland, and the Laurasian landmass. Prior to the pre-Cambrian period, indications are that there was only one huge landmass and only one ocean, which are given the names of Pangaea and Panthalasa, respectively. Therefore, it may be said that the number of continents has grown from 1 through 2 to the present-day figure of seven.

SIZE AND DISPOSITION OF CONTINENTS

It is very interesting to consider the location of the continents alongside with their sizes (surficial extent). It is seen that

(i) North and South American continents are close to one another, their sizes in sq.km. being, 23117691, and 17730345, respectively,

(ii) Europe and Asia, infact constitute only one landmass, while Africa is slightly separated from it by the Mediterranean ocean. Their sizes in sq.km. are 27281142, 27391512, and 29981679 respectively,
(iii) Australia and New Zealand are close to each other, their sizes in sq.km. are 7686810 and 268676 respectively, and
(iv) Greenland is 2175600 sq.km. in size.

The feature to be noted is that the American continents taken together measure only 40848036 sq.km., and are far away from Europe, Asia and Africa, constituting a contiguous landmass. This latter landmass is as big as 84654333 sq.km., and is double in size, in comparison with the American continents taken together.

Considering the sizes of the continents, it is quite apparent that there are only two big continents, the sizes of Australia, New Zealand and the Greenland being very small. These features are depicted in Fig. 12.

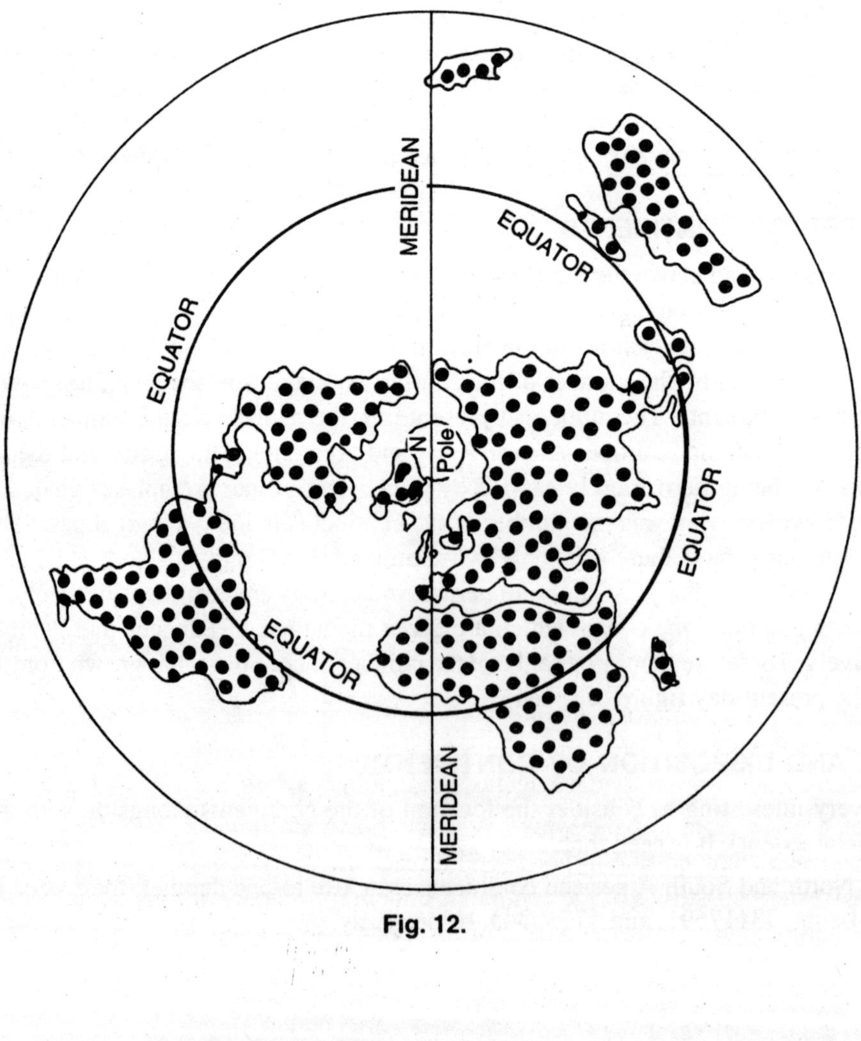

Fig. 12.

Unequal disposition of landmasses across the *Zero Meridean*. Note that landmasses are more extensive on the right hand side of the zero meridean (entire Europe, Asia, major part of Africa, Madagaskar, Australia and New Zealand), while only North America, South America and Greenland occur on the left hand side of the zero meridean. Landmass on the right hand side is double in extent than that on the left hand side. Has this feature any genetic significance?

A careful scrutiny of Fig. 12 reveals the unequal disposition of the two continents with respect to the line of Meridian of the Earth. Only Greenland and the two American continents are situated to the western side of the line of the Meridian, while the giant sized continent is spread to the eastern side of the line of the Meridian. Is there any significance to the inequality in the sizes of the two major continents, and their disposition across the line of the Meridian? Taking into consideration the great contrast in the sizes of the two landmasses, it is suggested that, instead of dividing the Earth into the northern and the southern hemispheres along the line of the Equator, it is more realistic to divide the Earth into the western and the eastern hemispheres, along the line of the Meridian.

SHAPES OF THE CONTINENTS

Let us now turn to the shapes of the continents. The surface of the Earth is described to be constituted of several continents and oceans. There are seven continents, the names of which are already given in the section dealing with the "names of the continents". Excepting Europe andAustralia, the shae of the other continents attracts attention. The two Americas, the Greenland, Africa and India possess a shape somewhat comparable to that of an isosceles triangle. Further, the northern coastline of these continents is quite broad but short in extent, while the eastern and the western coastlines are very long but taper almost into a point, in the southern direction. Thus the northern, the eastern and the western coastlines built up the isosceles shape of the said five continents. Does this feature of shape tentamount to an "accident" in the evolution of the continents, or it has some other significance? Were these continents "born so", or these have been later "shaped so" by the erosive action of the geological agents?

Geological agents like the rivers, the glaciers and the wind, erode the landmasses, and produce various shaped geomorphic forms like the Grand Canyon of Colorado river, the Mesa and Butte hills, the Trap scenery, the Cocks-Comb structure, the Tor scenery and so on. The geological agents denude the landmasses, reduce the height of the elevated parts, and practically reduce them to the sea level. However the said agents do not shape the coastlines of the continents. *The line of contact of the oceanic and the sea water with the landmass, defines and controls the shape of the coastline*. The oceanic and the sea waves continuously and constantly dash against the continents.

The erosive action of the waves is in the horizontal plane. This action can be best described as a "Saw cutting in horizontal direction". Considering the fact that the eastern and the western coastlines of the continents under discussion, are considerably longer than the northern and the southern coastlines, it sounds logical to construe that the oceanic and the sea waves moving in the E-W direction, were more active than the waves moving in other directions. As a result, these continents have developed longer, and N-S trending coastlines. The erosive action of the E-W moving waves is expected to produce northern and southern coastlines of the North and the South Americas, of nearly equal lengths. But the fact is that these two continents do not possess any coastline at all, at their southern borders! Instead, a "tail shaped" form is developed. There is one more feature of interest that needs a mention. The two American continents are separated in a puzzling manner. *Whereas the southern portion of North America has tapered into a "tail", the northern coastline of South America is widest in that region.* The isosceles triangular shape of the continents throws up the problem of origin. Assuming that the E-W flowing waves were more powerful and active, then what led these continents to develop a "tailshaped" form only for the southern coast? What is more mysterious is that these waves have acted differentially at the meeting point of the North and the South American continents. How is it that the erosive action was strong at the *southern coastline* of the *North America*, but at the same time not so for the *northern coastline* of the *South America*? It is thus quite apparent that unusual situation prevails in the region where the North and the South American continents meet.

The waves of the Pacific ocean have strongly affected the western coastlines of the North and the South America. The sea and the oceanic waves created in the Atlantic ocean have affected more strongly the eastern coastlines of the North and South Americas, and the Greenland, the west coastline of Africa and the Greenland, but not so much the west coastline of Europe. The oceanic waves appear to have been assisted by some structural features like E-W trending wide shear zone, or a major fault in the region marking an almost detachment of the North and South Americas. Likewise the San Andreas fault bordering the western part of the two American continents has assisted the waves to shape the western coastlines to trend in a general N-S direction over a very long distance. The waves of the Indian ocean have affected the eastern coastlines of Africa, and the eastern and the western coastlines of India. At this juncture it is necessary to remember that the continents have also undergone extensive drifting. It is, therefore, possible that the process of drifting might have contributed in shaping the different continents. Most perplexing feature about the shape of the continents is that all of them have a very broad northern coast, and a narrow southern tip. The determination of the modus operandi of shaping the continents, therefore, deserves investigation.

SEQUENCE OF FORMATION OF CONTINENTS AND OCEANS

Regarding which event was first - the oceans or the continents - the view of Dott and Batten (op. cit.) are very much useful, and are thought-provoking. They write, ""Elements evolve from hydrogen through a series of complex thermonuclear reactions in stars. Sea water is formed by oxidation of free hydrogen. The greatest single problem facing any hypothesis of development of the atmosphere is to explain the abundance of free oxygen, which is missing from Jupiter and the meteorites. The most refractory elements Ca, Al, and Ti could condense into chemical compounds in the range of 1500-1300°C. Elements such as Fe, Ni, Co, Mg and Si would condense and form compounds in the range of 1300-1000°C. Remaining elements would condense between 1000-300°C. Water would condense below 100°C"

Thus the modus operandi of formation of elements clearly indicates that first the minerals (rocks) were formed, followed by water. *The sequence of formation was, therefore, that of the crustal rocks first which formed the continents.* It is thus quite certain that the oceans came into existence only after the formation of the landmasses (continents).

CHAPTER 6

ORIGIN OF CONTINENTS, OCEANS AND PLATE TECTONICS

INTRODUCTION

The shapes, sizes (surficial extent), disposition and the number of the continents, are closely related to the origin of the Earth, especially to the formation of the crustal part of the Earth. The opinion has swung from the "Permanancy" to that of "Drifting" of the continents and the oceans. The building activity in the crustal part has not ceased even today, which is evidenced by the occurrence of the earthquakes, and the volcanic erruptions. It is worthwhile to note the various opinions held by the geoscientists about the evolution of the continents and the oceans. J. Tuzo Wilson (quoted by Schlee, op. cit. p. 17) wrote "if Earthscientists have been trying to fit the history of Earth into the framework of a rigid and fixed pattern of continents, then it is not surprising that it has been impossible to answer the major questions. It is not our methods nor our observations that have been wrong, but our whole attitude". Evolution of the continents has been studied and discussed by several geoscientists, noteworthy being, A. Sneider (1858), F.B. Taylor (1908), A. Wegener (1910, 1966), A.L. Du Toit (1937), T.H. Holland (1941), A. Holmes (1960), S. Schlee (1975), R.G. Park (1983), K.C. Condie (1984) and so on. Schlee (op. cit.) states" a study of fossil magnetism suggested that the subcontinent (India) had travelled well over 4000 miles from a position south of equator to its present position in the northern hemisphere (p. 340). Crust of the Earth consists of 6 or more rigid plates. These are continuously created at one edge and destroyed at another (p. 362). Holmes (1960) has put forth three primary units namely, (i) continents of Laurasia group, (ii) the continents of Gondwana group, and (iii) the Pacific ocean, and wrote (1981, p. 487). "To what extent these primary units and their arrangement have been stable or otherwise during geological time is one of the fundamental problems of geology". Dana (1846) held the view that the continents and the oceans have never changed places. Sneider (op. cit.) opined that during the late Carboniferous period, all the continents were together and constituted only one gigantic landmass, which was later named as Pangaea. Condie (op. cit., pp. 248, 249) states, "A few supercontinents existed on the Earth during most of the pre-Cambrian, major fragmentation began about 700 m.y. ago". Park (op. cit., p. 113) divides Earth's surface into six major plates, namely Eurasian, American, African, Indo-Australian, Arctic and Pacific.

Thus it is apparent that opinions and information varies about the "number of continents and oceans, the disposition of the continents, size and shape in three dimensions of the continents, nature of contact of the continents with one another, and the modus operandi about the movement of the continents". It is, therefore, thought worthwhile to elaborate and discuss the complicated and perplexing problem of evolution of the continents, in the background of the hypothesis of "Plate Tectonics".

A PERSPECTIVE APPROACH TO THE CONTINENTS AND THE OCEANS

The continents and the oceans are normally viewed from the standpoint of these as constituting the surface of the Earth, and building up the crust of the Earth. The extent of the continents in depth (thickness), and in the horizontal direction (length and breadth), is also required to be considered, in order to describe the size and the shape of the continents. Whereas the depth could be ascertained, the values of length and breadth fluctuate, because these parameters depend upon the position of the sea level. In fact the demarcation of the continents depends upon the expanse of seas and oceans. The continents appear to be separated from one another, but in fact these form a continuous land scanning the entire surface of the Earth. This feature can be readily appreciated in Figs. 9A, B, C. The dimension of length and the breadth of the continents is controlled by the volume of the oceanic and the sea water. The continents should, therefore, be visualised as big and small landmasses apparently separated from one another by the oceans and the seas. It also sounds improper to arbitrarily dissect the land surface into 5, 6 or more number of continents, and delineate them with boundaries. It is felt that in actuality boundaries do not exist at all, because the surface of the landmasses (continents) is unbroken and is a continuous phenomenon. What is implied by this description is that the continents are not at all separated from one another, and that there are no gaps or open spaces in between them. The entire landsurface is a continuous entity. Park (op. cit., p. 113) holds the view that "plate boundaries may or may not correspond to continental margins". This view further complicates the definition of a "continent", indicating thereby that in reality boundary for the continents is not existing.

As described in Chapter 5, the surface of the Earth being rugged one, the depressions and the low lying areas got filled up by the rain water (when the Earth was born) giving rise to the oceans and the seas. This resulted in the development of continents from such of the landmasses that had higher elevations. The first formed continent or continents, must have originated by such a mechanism. When the outer part of the Earth cooled from its initial molten state, it might have given rise to a corrugated surface, some of which could have been more long than wide. This might have produced the ridges. As noted in Chapter 5, oceans were formed only after the production of the rugged outer surface of the Earth. The first showers of the rains must have filled up the depressions and given rise to numerous small sized water bodies (seas and lakes). With the recurrence of rains, the numerous water bodies (seas and lakes) must have coalesced and ultimately given rise to extensive stretches of water bodies i.e., the present-day seas and oceans. It is also probable that the surface of the Earth when formed, could have been a smooth surface, without any elevations and depressions at all! In that event, there must have been only expance of water over the entire surface of the Earth. *Water and water everywhere, and no landmass at all*!

In the development of the continents and the oceans, it is very tempting to involve the origin of the Moon, the strong point being its proximity to, and its act of rotation around the Earth. The crustal rocks of the Moon are basic in composition, and the core is a mere 200-400 km in thickness (Table 4). It is not known whether the core is made up of nickel and iron, like that of the Earth. On the other hand, the crustal rocks of the Earth are acidic in composition (Table 3), the core is 5200 km thick (Table 2), and is composed of nickel and iron (Table 3). What is more important to observe is that the lunar crust is older than any rock yet dated from the Earth as noted by Condie (op. cit., p. 18). The hypothesis of the formation of the oceanic depression of the Pacific, due to the detachment of a large chunk from the crustal part of the Earth, and giving rise to the Moon, receives a setback. However, notwithstanding any other mechanism that could result in a gigantic sized (165,384,000 sq.km.) scar (of the Pacific ocean), probability of detachment of enormous landmass from the crustal part of the Earth, can not be outrightly rejected either.

A PERSPECTIVE APPROACH TO PLATE TECTONICS

The peripheral part of the Earth is designated as the crust, and it is split into (i) continent, (ii) continental slope, and (iii) floor of the oceans. The thickness in km of these parts according to Condie (op. cit., pp. 71-72) is 30-80, 15-30 and 5-15, respectively. He has called the continental slope by the term "transitional crust". His description about the "plates" (continents) leads to confusion. Condie writes (op. cit., p. 10), "a mosaic of plates bounded by oceanic ridges, subduction zones and transform faults". Does it mean that each plate is surrounded by these features? Could a plate not exist in contact with another one, without a ridge, a subduction zone, or a transform fault? Moreover faulting can take place only if the landmass (plate/continent) is a continuous body. A landmass must exist first, only then ridges, subduction zone or faulting could take place. Schlee writes (op. cit., p. 362), "The boundaries of these plates are mostly easily located by mapping earthquake centres. The boundary between the American and the Pacific plates is located in part along the San Andreas fault in California". It does not sound reasonable to render the San Andreas fault as the boundary between the American and the Pacific plates, because this again indicates that prior to faulting, the landmass was continuous, and not as American and Pacific plates. The feature to be underlined is that the continents do not possess any boundary around them, the implied meaning being that the coastline defines the limit of any continent and the development of a coastline depends upon the level of oceanic water. Another feature deserves attention. What is the cause of the "thinness" of the floor of the seas and the oceans? Is it in any way related to the origin of the Moon, which is believed to have sprung away from the Earth, and creating the extensive depression of the Pacific ocean?

Let us now analyse the extension of the continents at depth. The Earth is made up of three shells, the terminology adopted being the crust, the mantle, and the core. The crust is further split into Sial, and Sima, on the basis of density. What is necessary to stress is that the sial merges imperceptibly into the sima. In other words it may be said that there is no rigid or a sharp boundary in between the sial and the sima, as well as that between the mantle and the core. In order to appreciate the varying thickness of the crustal (sial) layers, Figs. 10A, B, C, D have been drawn.

The continents are not only not uniformly thick, but also lack definite boundaries. Of late, even the term "continent" is dropped, and instead the crust of the Earth is considered to be made up of "plates". Thus Condie (op. cit., p. 1) states, "surface of the Earth is composed of rigid plates rest on hotter, denser layer, known as asthenosphere". According to Park (op. cit, pp. 113, 114), a layer of abnormally low seismic velocities exists at a depth of around 100-150 km, which is called "low velocity zone". There is a decrease in density and viscosity resulting in a more ductile layer, which exists between the plates (continents/crust) and the mantle. Park also suggests that the bottom part of the continent (called as plate) is not distinctly divided or separated from the underlying mantle. Radhakrishna and Vaidyanathan write (1977, p. 27), "The plate tectonic theory envisages the existence of a dozen rigid plates not fixed in their position but are floating over partially melted material. These plates are believed to be constantly in motion, the rate of movement being very slow". Based on a perspective approach to the three dimensional shape of the continents, Figs. 13A, B have been drawn. In Fig. 13A, South American and African continents, and the Atlantic ocean in between them, have been chosen as an illustration to depict the behaviour of the continents at depth, and along the continental slopes, while in Fig. 13B, the continents shown are hypothetical ones, which are separated by the Atlantic ocean.

Figs. 13A and B clearly display the three dimensional shape of the two continents, and that of the oceanic floor (the continental slopes). Whereas the continental crust is considerably thick, but laterally i.e., along the slope (what has been designated as the "transitional crust" by Condie (op. cit., pp. 71-72), it is considerably thin. To mention the average values, these are 33 and 6 km corresponding to the continental crust and the continental slope, respectively. The continental crust is thicker but small in its lateral extent (the extent is only 149 million sq.km.). The oceanic floor is thin, but it is very extensive (it being 361 million sq.km.). Therefore the shape of the continents resembles that of an irregular block, and not that of a plate. Further, the feature to be realised is that the crustal blocks of the continents,

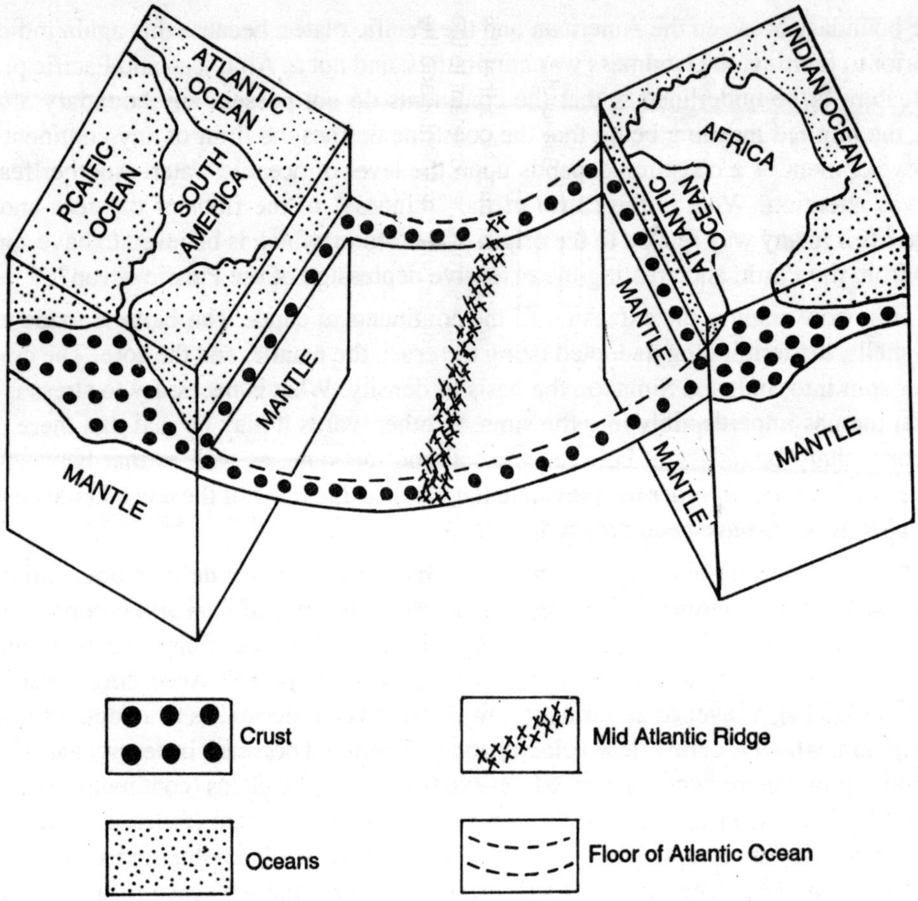

Fig. 13A. Projection of the crust beneath South America, Africa, Pacific, Atlantic and Indian oceans. Note that the crust is considerably thick beneath the continents, and it is extremely thin beneath the floor of the oceans.

and the crustal slabs of the oceans, constitute contiguous entities, and therefore, these should not be deemed and projected as independent bodies.

FEATURES OF PLATES

In order to derive the inter-relations between the several plates and their compositions, Fig. 14 has been drawn, based on Fig. 14.15 of Park (op. cit., p. 119). Information gained from the scrutiny of Fig. 14 has been set out in Table 7.

Study of Fig. 14 and Table 7 reveals that the plates differ considerably in their surficial (length and breadth) extent, the biggest being the American plate, which practically extends from the North Pole and almost reaches the South Pole. Smallest is the Indian plate. Features set out in Table 7 indicate that :

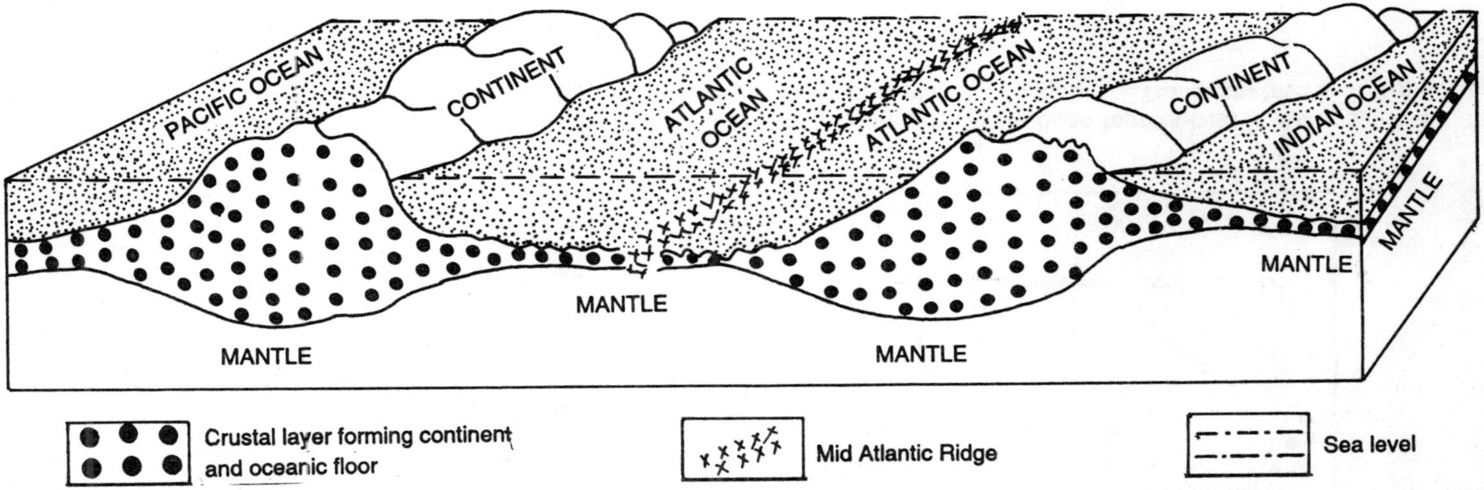

Fig. 13B. Three-dimensional features of the continents and the oceans. Shape of the continents is hypothetical one. Note that the continental crust is very much thicker than that of the oceanic floor. Also note that the crustal layer continues from the continent to the adjacent one, via the continental slope and the oceanic floor, without any gap or break in between. Distinction into continental and the oceanic plate is therefore arbitrary.

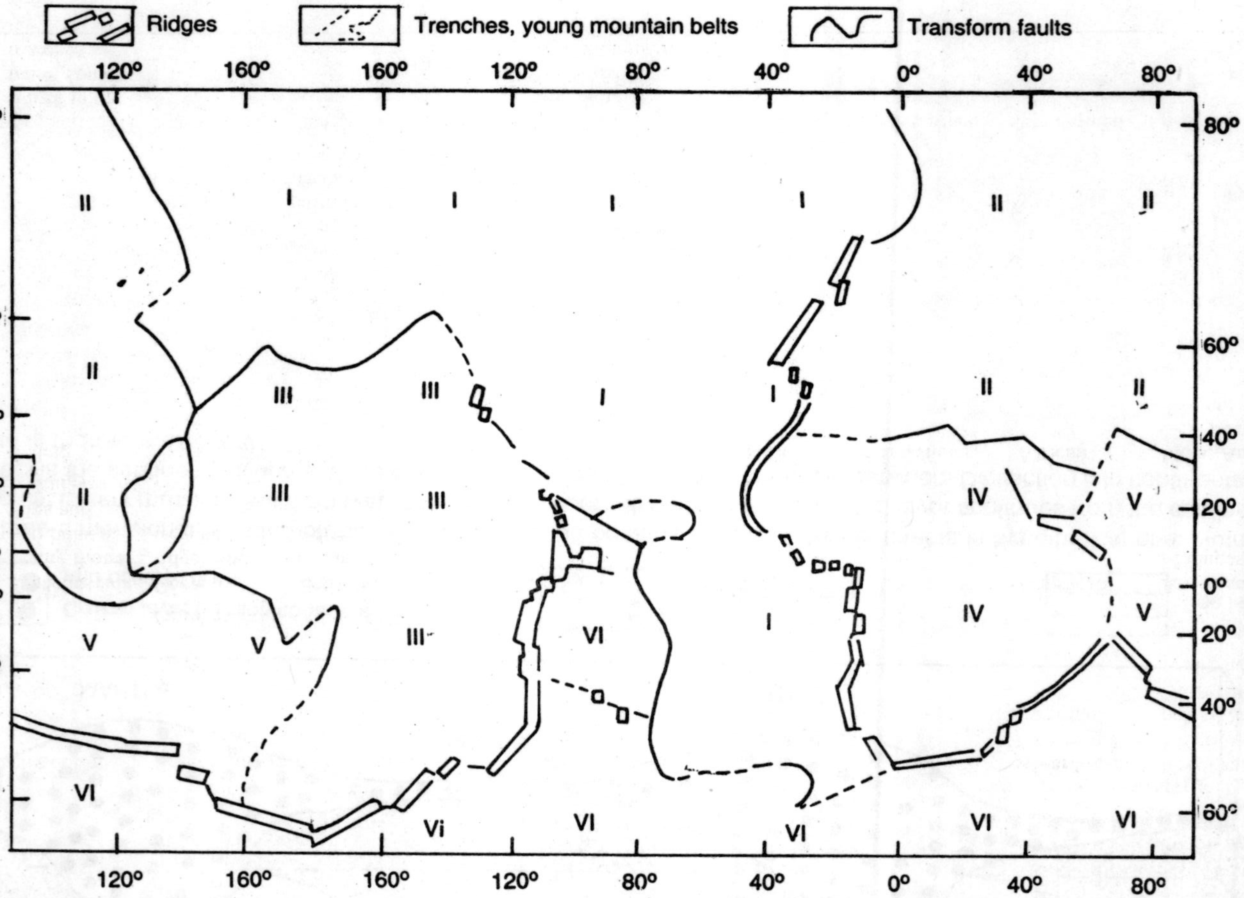

Fig. 14. Disposition of plates (after Fig. 14.15 of R.G. Park, 1989, only boundaries have been reproduced). I. American plate, II. Eurasian plate, III. Pacific plate, IV. African plate, V. Indian plate, VI. Antarctic plate. Ridges, Transform faults Trenches, young mountain belts.

Table 7. Features of (crustal) plates (derived from Fig. 14.15 of Park, 1983, p. 119

Name of plate	*Composition of boundary of the plate*				*Name of surrounding plates*				*Continents composing the plate*
	North	*East*	*South*	*West*	*North*	*East*	*South*	*West*	
Pacific	Oceanic trenches	Transform faults; San Andreas fault; Oceanic trenches	Oceanic ridges	Oceanic trenches in the main; Transform faults	American	Partly American partly Antarctican	Antarctican	Partly Indian, partly Eurasian	A few tiny Islands; rest covered by the Pacific Ocean
American	Not defined	Entirely made up of mid-Atlantic ridges; a small stretch of of ocean trenches	Transform faults and oceanic trenches	San Andreas fault; Rockies and Andes mountains	Not defined	Eurasian and African	Antarctican	Antarctican, Pacific and Eurasian	North and South Americas, Eastern part of Siberia, Greenland, part of Arctic Island
African	Trenches and young mountain belts of Alps and Himalayas	Oceanic ridges and trenches	Entirely oceanic ridges	Part of mid-Atlantic ridges	Eurasian	Indian	Antarctican	American	Africa and Madagascar
Indian	Young mountain belt of Himalayas; Oceanic trenches	Transform faults; Oceanic ridges	Entirely oceanic ridges	Oceanic ridges; Transform faults	Eurasian	Pacific	Antartican	African	India, Australia, New Zealand, Sri Lanka
Eurasian	Not defined	Oceanic trenches; young mountains; transform faults	Young mountain belt of Alps and Himalayas; Oceanic trenches	Mid-Atlantic ridges; oceanic trenches	Not defined	Partly American, partly Pacific	African and Indian	American	England, Central Europe, Russia, Part of Arctic Island
Antarctican	Oceanic ridges in the main; some trenches	Not defined	Not defined	Not defined	Indian, Pacific, African and American	Not defined	Not defined	Not defined	Antarctic Island

1. plates are demarcated from one another on the basis of the presence of the trenches, the ridges, the transform faults, and the young mountain ranges,
2. plate boundaries are dominantly comprised of ridges, young mountain belts (the Rockies, the Andes, the Alpine-Himalayan ranges) followed by the trenches and transform faults (the northern and the southern border of the Pacific plate). There are no gnawing gaps in between any plates,
3. main yardstick used in delineating the plates, is the existence of mountain belts, as is exemplified by the entire western boundary of the American plate, which is made up of the Rockies and the Andes mountains. The entire southern boundary of the Eurasian plate also belongs to a similar category, it being made up of the Alpine-Himalayan ranges,
4. there is only one oceanic plate, it being the Pacific plate. The rest comprise of "continental plates",
5. Antarctic plate has been projected, though it is entirely capped by thick sheet of ice, but the Arctic plate has not been contemplated. Probably the Arctic region is regarded to be northern extension of the American and the Eurasian plates.

In order to appreciate the actual inter-boundary relations between the plates, two hypothetical vertical sections in E-W direction, have been drawn between (i) the North America and the Eurasian plate, and (ii) South America and the African plate. These are presented in Figs. 15A, B. In Fig. 15C, a seismic section along British Isles, North Germany, Europe, Asia and Japan, is shown.

Scrutiny of Figs. 15A, B, C clearly reveal that the continental and the oceanic plates form a contiguous body, and are without any boundary in between them. That boundaries do not exist in between plates is clearly observable in Fig. 15C, which is a seismic section, and not any hypothetical projection. Fig. 15C also demonstrates that the thickness of the continental and the oceanic plates, differs considerably.

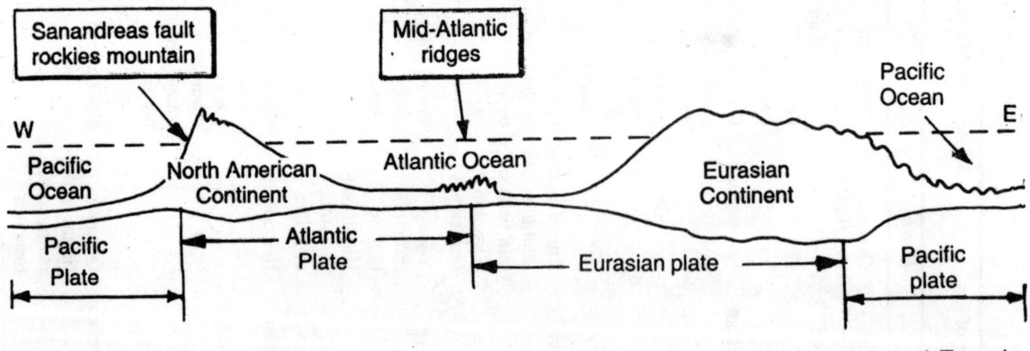

Fig. 15A. Hypothetical vertical section in E-W direction between North America and Eurasia. Note that the crustal layer is very thin beneath the oceanic floor.

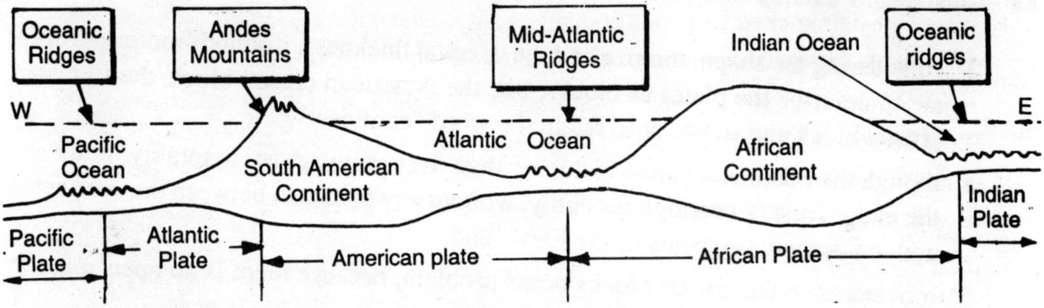

Fig. 15B. Hypothetical vertical section in E-W direction between South America and Africa. Note that the crustal layer is very thin beneath the oceanic floor.

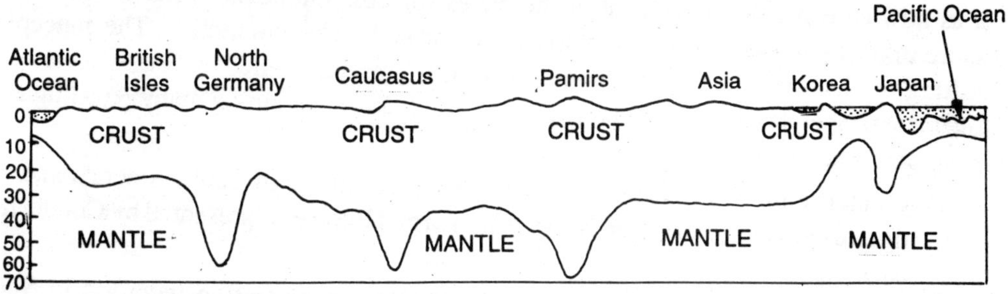

Fig. 15C. (After Fig. 699, of Arthur Holmes, 1966). Note that the crustal layer is extremely thin (about 10 km) under the oceans, while it is 30-40 km thick under the continents.

FALACY OF PLATES

Continents are at least visible and hence acceptable, as long as the oceans and the seas exist. This is not true about the plates. The remarks of Park (op. cit., p. 113) that the plate boundary need not coincide with that of the continent, raises doubts about the very existence of the plates. The San Andreas fault, the mid-Atlantic ridges, the Alpine Himalayan mountain ranges, are made use of, in creating the Pacific, the Eurasian and the Indian plates. Therefore, these plates were not existing prior to the development of the said features, and the land surface was a contiguous entity. The concept of plates and their tectonics, instead of finding a solution to the perplexing hypothesis of continental drift, it creates confusion, because the plates are not the outcome of any natural geological event.

From the foregoing account of the crustal plates, it is seen that :

1. oceanic ridges, mountain belts, trenches, faults and the continents, have all originated only after the episode of the formation of the crustal shell from the primordial molten mass of the Earth. *This leads to infer that the crust of the Earth did not consist of plates, to begin with,*
2. the continental plates are much thicker than the oceanic plates. There is only one oceanic plate, rest are all continental plates,

3. considering the shape, the size and the unequal thickness, it sounds more appropriate to describe the plates as blocks, like the American crustal block, the Eurasian crustal block and so on, or to retain the word "continent",
4. though the continental crust is thicker than the oceanic crust, in totality, however, the entire crust is a contiguous entity, without any breaks in between the continents and the oceanic segments of the Earth, and
5. movement of the crustal blocks poses problem, because there is no open space in between them.

MECHANISM OF PLATE MOVEMENT

Let us now consider the movement of the plates (blocks/continents). Prior to that, let us enquire into why it was felt necessary to set the crustal blocks into motion. The concept of "Drifting of Continents" is a fore-runner of what is presently called as "Plate Tectonics". Movement of the plates (continents) has been initiated for explaining many puzzling geological phenomena like :

(i) geometric fit of the eastern and the western coastlines of South America and Africa, respectively, and even North America and Europe as presented by Condie (op. cit.), Fig. 16,

(ii) palaeo-climatical features (Permo-Carboniferous glaciation recorded in South America, Africa, India and Australia),

(iii) plant fossils preserved in the Coal Measures of Europe, being identical with those of North American Coal Measures,

(iv) South American Triassic reptiles also found in the Triassic of South Africa, these two landmasses that are at present 4750 miles apart, the Atlantic ocean separating the two continents,

(v) discovery of the 20th Century that "Man originated in Africa and later migrated to different landmasses (continents) based on the finding of similarity of DNA characters of 50,000 year old specimens with that of man of today from Africa,

(vi) palaeontologists working in Kenya have found the fossilised remains of a human face and skull thought to be 3.5 m.y. old which will shed new light on man's evolution, the journal *Nature* reported. Mr. Museums of Kenya and colleagues found a battered but almost complete skull and face of an entirely new breed of early human (information collected from the *Hindu, India's National Newspaper*, dated 23.3.2001),

(vii) According to Dr. Spencer Wells, all human DNA coalesces to a single ancestor living in Africa around 60,000 years ago. The original African closest living relatives now live in Namibia, Botswana and the northern part of South Africa. Around 50,000-60,000 B.C., a group of Africans started walking. In those days sea level were far lower than they are now and there was virtually unbroken coastline all th

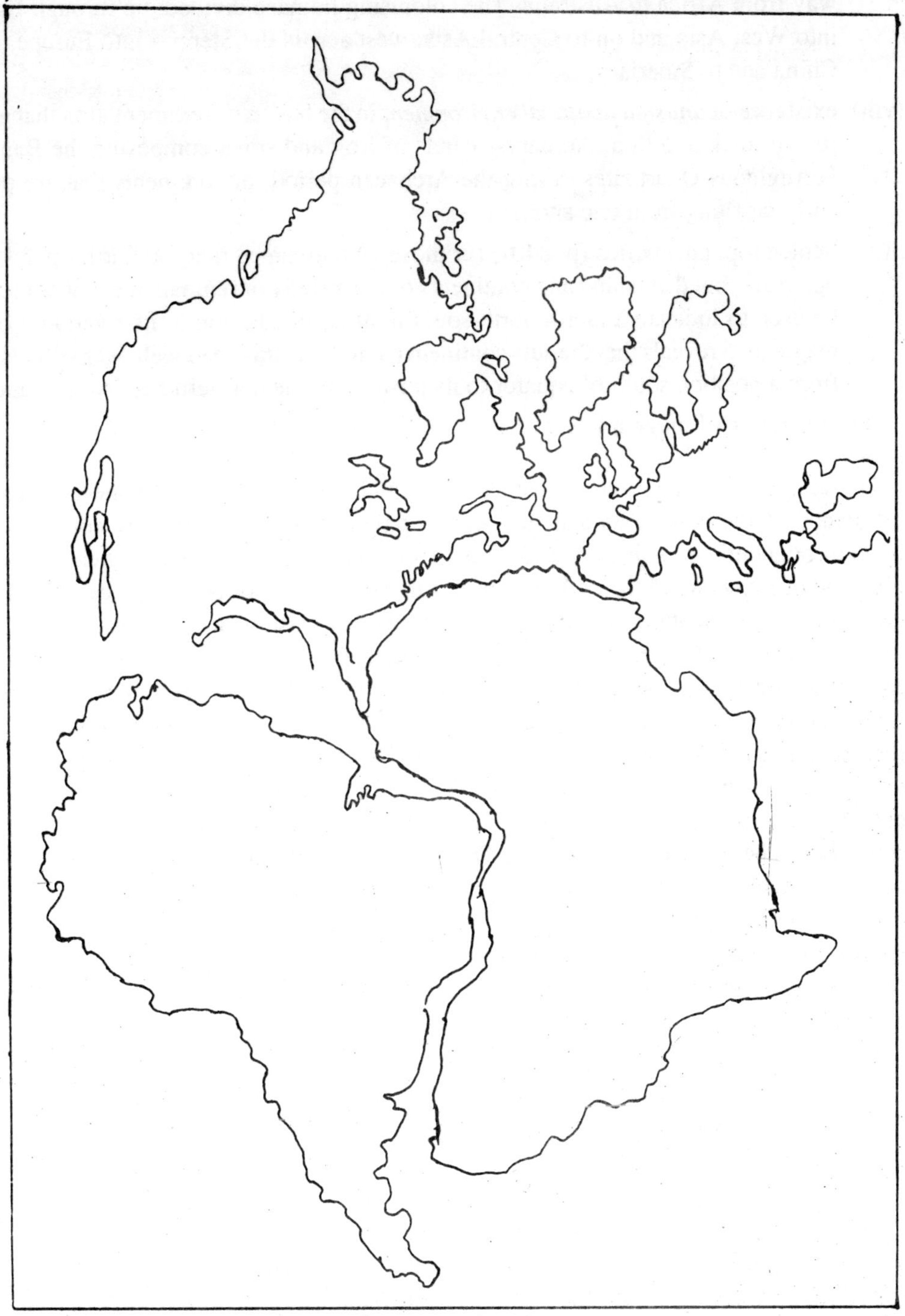

Fig. 16. (After Fig. 8.1 of K.C. Condie, 1984). Computerized least square fit at 500 fathoms around the Atlantic ocean (from Bullard et al., 1965).

way from Africa to Australia. The colonising humans divided; up through Egypt into West Asia and on to Central Asia; west across the Steppes into Europe, into China and to Siberia,

(viii) existence of *unusual identical environment* in the basins of sedimentation that gave rise to thick and thin alternating bands of iron and silica composing the Banded Ferruginous Quartizites during the Archaean period, in continents that are presently far flung from one another,

(ix) Schlee (op. cit.) writes (p. 340) "for those who assumed that the Earth's poles did not move, but that island and continents did, the study of remnant magnetism could be used to indicate a land's north-south motion, in addition to its rotation. Fossil magnetism reveals that the sub-continent (India) had travelled well over 4000 miles from a position south of equator to its position in the northern hemisphere, and

(x) generation of magma in space and time.

The above noted geological features, except item No. (X), shake the concept of permanancy of the continents and the oceans, paving way for setting the continents in motion throughout the geological past. It is firmly held that the movement being exceedingly slow of a few cm (5-10 cm) annually, it therefore, can not be felt by us. Out of the above noted several features, a near perfect fitment of the eastern and the western coasts of South America and Africa, respectively, give credence to the idea that these continents were joined together in the past. Based on this fitment, the continents have been reassembled into two super-continents called (i) the Gondwanaland (South America, Africa, India, Australia and Antarctica), and (ii) Laurasia (North America, Greenland, Europe and Arctic). Some of the geological features even lead to huddle all the continents into one giant landmass, as had been done by Sneider way back in 1858. It is called as Pangaea, the singular ocean being named Panthalasa. The concept of Permanancy of continents, fails to tender satisfactory explanation for the similarity of many geological features possessed by continents that are far flung from one another, at present. The concept of drifting (movement) of continents on the other hand, faces the problem of reckoning a powerful motive force that can set the continents in motion, either towards or away from one another. Such a motion has to be in a horizontal direction. Obstacles in such a movement are two fold :

(i) every continent is in juxtaposition with one another, meaning that no gap or space exists in between them, and

(ii) generating a force capable (powerful) of setting the large sized continents in motion.

Regarding the first obstacle, it is to be remembered that all continents together constitute a contiguous crustal part of the Earth. Granting for a moment that a continent is to be set in motion, its contact with the continents surrounding it, as well as with the mantle forming its bottom, is required to be severed. Such a detached continent still can not move towards or away from the other, because gaps do not exist between the continents. The modus operandi of movement, is therefore, complicated and very puzzling one. In order to tender explanation for numerous puzzling geological features :

(i) the continents are termed as plates,

(ii) drifting is substituted by the term "plate tectonics", and

(iii) drifting is regarded to be caused by sea floor spreading.

SEA FLOOR SPREADING - A TERM OF CONFUSION

In the concept of "Plate Tectonics", "Sea floor spreading" has assumed great importance. This term however creates an impression that, the "*crustal layer forming the seas and the oceans, alone is set in motion, and it causes the drifting of the continental plates*".

Let us now try to understand as to what exactly is meant by the expression "sea floor spreading". Park (op. cit., p. 110) writes, "The hypothesis of sea floor spreading put forth in 1962 by Hess and Dietz, linked continental and oceanic areas in the same general process, likened to a conveyor belt, in which slabs of crust were transported laterally, carrying the continents with them. The two hypotheses of continental drift and sea floor spreading, became combined to form the nucleus of the new hypothesis of plate tectonics". The term "sea floor" spreading, became combined to form the nucleus of the new hypothesis of plate tectonics". The term "sea floor" thus appears to be used as a synonym for "oceanic plate", because the crust of the Earth has been regarded to be made of continental and oceanic plates. Another confusion is regarding as to why it is designated as "sea floor spreading" and not as "oceanic plate spreading". Is sea floor different from the oceanic plate? It also creates an impression that the drifting of the continents is the outcome of the spreading of the sea floor. When the sea floor spreads, its thickness should get reduced. That is to say, that it should become thinner, than what it was previously. The thickness of the oceanic plate (floor) is about 5-7 km, while that of the continental plate is 130-150 km. Can a thin oceanic plate of 5-7 km thickness, pull or push large sized and much thicker continents, laterally? Instead of designating it as a process of "spreading", the word "movement" is more apt, because then the entire thickness of the oceanic plate will be involved in the process of movement of the continents. Schlee (op. cit., p. 362) writes, "most recently the theories of sea floor spreading and continental drift have been subsumed within the concept of plate or global tectonics". Even in this description, there is no clarity about the term "sea floor spreading".

Condie (op. cit. pp. 125-126) writes, "New lithosphere is generated at oceanic ridges, spreads laterally and returns to the mantle in subduction zone. Relative motions of lithospheric plates can be described in terms of a pole of rotation passing through the earth's centre. Two models for the forces driving plates are the viscous drag and buoyancy models. The viscous drag model involves coupling between the lithosphere and the asthenosphere, as mantle convection currents spread laterally. In the buoyancy model, plate movement is produced by negative buoyancy in subduction zones, pushing at oceanic ridges and gravity sliding of plates away from the ridges". This account too does not clarify the term "sea floor spreading".

Fig. 17 shows that the sea floor (lithosphere) has not spread, but it has merely moved below and down to deeper level. The expression "sea floor spreading" is thus misleading, because the act of spreading should have affected the thickness, but the thickness of the sea floor (lithosphere) has remained unchanged. However Fig. 17 does not give any idea about the disposition of the sea floor (lithosphere) prior to attaining the features shown in the said figure (Fig. 17). The positions of the sea and the continent with respect to the lithosphere also are needed to be marked in that figure (Fig. 17). Such a figure then assumes a realistic outlook. Condie (op. cit. p. 16) writes, "magmatic events may record tensional/compressional stresses in the crust, or they may be completely unrelated to any known crustal deformation. Their interpretation in terms of tectonics is critically dependent upon field relations". It is thus quite evident that the term "continental drift" is more befitting than the expression "sea floor spreading".

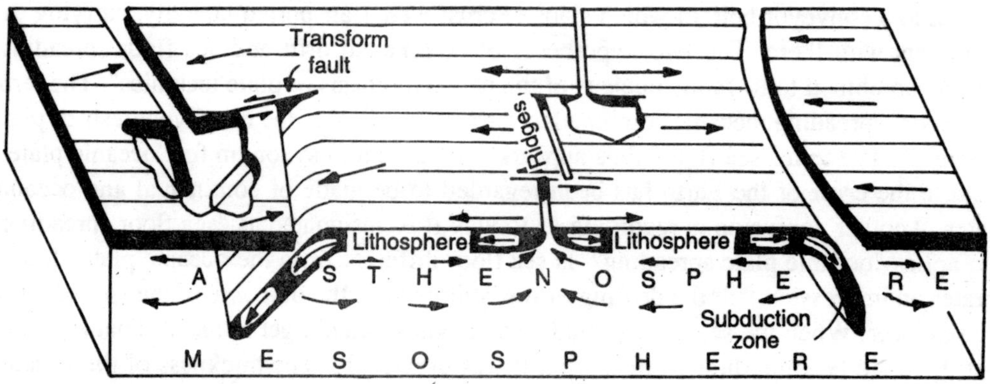

Fig. 17. Major features of sea floor spreading (after Fig. 1.1 of K.C. Condie, 1984).

SOURCE OF MOTIVE FORCE FOR DRIFTING

The motive force that could set the continents in motion is yet another debatable issue. It centers around two aspects :

(i) source of the force, and

(ii) its ability to drift the large sized continents over long distances.

Holmes (1960) introduced the concept of convection currents generated in the mantle underneath the continents. The mantle melts due to the heat generated during the disintegration of the radioactive minerals like pitchblende, uraninite, monazite and so on. Movement of the plate is effected by the uprising currents. Condie (op. cit., p. 140) writes, "Major sites of magma generation are at the convergent (subduction zone) and divergent (oceanic ridges) plate boundaries, and appear to be related to compressive and tensional stresses, respectively". Does it mean that the movement of plate generates magma? But then what causes the movement of the plate? Further, what causes compression or tension in the crust of the Earth? It appears that in order to explain the generation of magma, resort is taken to "plate tectonics". Even then, which event is the initial one, (a) generation of magma, (b) plate movement, or (c) creation of compression and/or tension in the crust?

Some geological events have remained unanswered like :

(a) why the Laurasian continent fragmented into 5 or more number of sub-continents during the late pre-Cambrian and Cambrian periods? What was the compulsion for the disruption? These sub-continents later collided again to reform the Laurasian continent. A mysterious event inedeed, because, why did the Laurasian continent in the first place, earlier fragment at all?

(b) the laurasian and the Gondwana continents unite to form Pangaea during the late Palaeozoic period,

(c) Pangaea splits during Mesozoic and Cenozoic periods, and

(d) the Gondwana land splits to form the continents of South America, Africa, Australia, Antarctica and India.

The above noted features lead us to ponder over the question namely, which is the initial episode amongst those noted below :

(i) splitting of continents,

(ii) reunion of continents,

(iii) the igneous activity, or

(iv) the plate movement?

Plate tectonics appears to be one amongst several other geological episodes. Condie (op. cit., p. 98) writes, "The causes of episodicity are : (i) changing convection patterns in the Earth with time, (ii) secular changes in the Earth's spin axis, (iii) episodic variations in shear melting of the mantle produced by tidal power and (iv) episodic magmatism and deformation related to plate tectonics". These features noted by Condie indicate that the concept of plate tectonics is to account for the origin of the various geological events (like mountain buiding, ridges, transform faults) that had occurred in the Earth, in the geological past periods.

A critical analysis of the concepts of the continental drift and the Plate Tectonics indicates that, both are practically one and the same, but for the substitutions of the words; Plate for Continent, Tectonics for Drift, and sea-floor spreading for Convection Currents. *Whereas earthquake, volcanic erruption, formation of plutonic igneous rocks, these are all geological events, the continental drift or the plate tectonics should be deemed as mere mechanisms or processes.* The continents are made up of rocks, and therefore, first the rocks are required to be formed. Thus unless the continents are built up, those can not be drifted. The rocks are formed through a natural event of crystallisation of magma. The role of "drifting or plate tectonics", is that of a "transporting agent". It does not build up any continent, but merely takes it from one place to another. Hence "Continental Drift, or the Plate Tectonics" can not be construed as forming any geological event or episode.

Magma generation, and generation of the convection current requires heat, and therefore, it is necessary to search for it. Condie (op. cit.) in this context notes that "heat generated at depth is primarily lost from the oceanic crust at the oceanic ridges, amounting to 75% of the total loss". There are two sources of heat namely :

(a) the hot interior core, and

(b) heat generated during the disintegration of the radioactive minerals.

Holmes (1981, p. 409) writes, "The whole of the heat lost from the earth could be supplied by one ounce of radium in every 1000 million tonnes of the material of the earth. The rocks themselves contain, on an average, fifty times as much. This startling result indicates that the radioactive elements must be largely concentrated in the crust - otherwise the earth would be much hotter than what it is, and we should not be here to discuss the matter - but it provides no reason for supposing that the material of the substratum can be absolutely devoid of these elements. Even minute traces would suffice to keep convection going".

MODUS OPERANDI OF CONTINENTAL DRIFT

There are many features which have remained unanswered while dealing with the continental drift and the plate tectonics. The first concern is about the dissipation of heat. Whatever may be the cause of generation of heat, why does it not bring about the metamorphic recrystallisation of the crustal rocks, or is not conducted through the crustal rocks, instead of giving birth to the convection currents? As per the observation of Condie (op. cit.), 70% of heat is lost from the oceanic crust, and that too, at the oceanic ridges. The oceanic crust is far more extensive (361 sq.km.) than the continental crust (149 sq.km.). Thus beneath the continents, only 30% heat is available, which has to initiate centers of upward rising convection currents beneath the continents. In fact, it is logical to expect convection currents underneath the oceanic crust, because maximum heat is lost from the oceanic crust. What is more astonishing is that these currents spread horizontally, as soon as they reach the bottom surface of sima, but do not metamorphose the rocks of sima?

Let us now concretely extend the process of drifting to some specific continent. The African plate is bounded on its west, south, east and north by the American, the Antarctican, the Indian, and the Eurasian plates, respectively. These boundaries comprise of oceanic ridges, oceanic trenches and the transform faults, and these features are presented in Fig. 18.

The geomorphic features around the African plate (Fig. 18) should be regarded as the effect of the process of drifting of the plates. To begin with the African plate must have been in juxta position with the four plates surrounding it. Under such a situation, the African plate has to move simultaneously in all directions, either towards or away from the plate surrounding it. How can such a movement take place, because there is no gap (space) available in between the plates? Such a movement is also required to bring about detachment of its bottom with the mantle. In addition to this, convection currents are required to be generated only below the African plate. Further, the African plate has to severe its con-

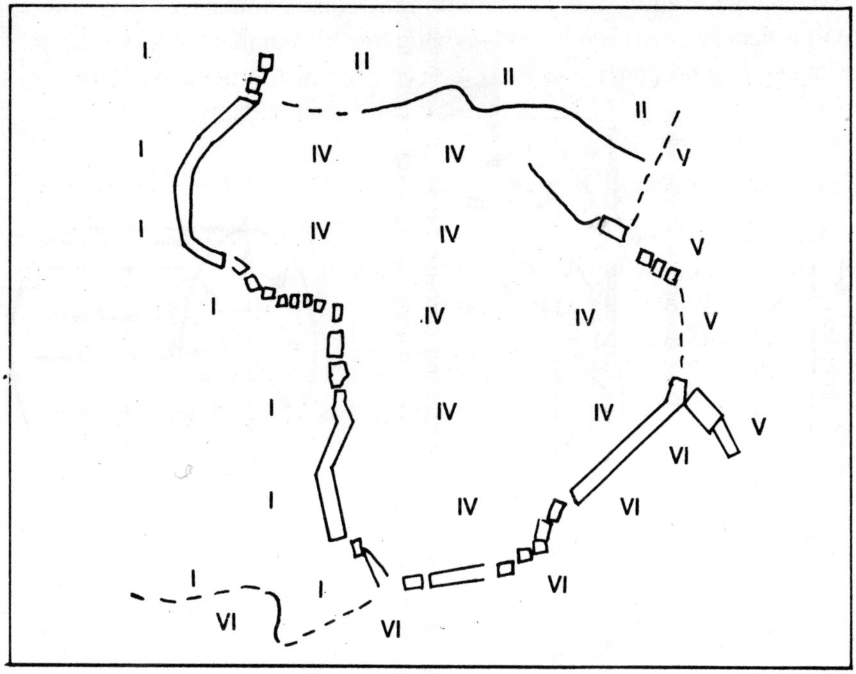

Fig. 18. (Drawn from Fig. 14) Boundary features of African plate. I. American plate, II. Eurasian plate. III. African plate, IV. Indian plate, V. Antarctic plate.

tact with the plates surrounding it. Thus the modus operandi of movement of plates assumes complexity, casting doubts about its feasibility. The exceedingly slow rate of drifting of plates over a few cms per year, renders the process less viable.

Whether it is plate tectonics or continental drift, elaboration is needed about the direction of movement or of drift. Holmes (1978) while dealing with plate tectonics has put forth a figure, wherein the directions of convection currents have been shown. It is reproduced here as Fig. 19.

Fig. 19 is a two-dimensional presentation. Another feature that deserves attention is that the rising currents stretch out the continent into two lobate parts, and the two parts drift away in opposite directions. It is, however, more realistic to project the rising convection currents to drift radially out in all directions, and not only into two opposite directions. This concept is presented in Figs. 20A, B.

A system of divergent convection currents generated underneath a continent (plate) is harnessed to set the landmass (plate or continent) into motion. The break up of the Gondawana and the Laurasian continents clearly demonstrates that the direction of movement of the convection currents was radial (Fig. 20B), and not merely in two opposite directions as presented in Fig. 20A. It is pertinent to recall here that the Gondawanaland must have split radially out, otherwise South America, Antarctica, Australia and India can not take up their present-day locations. This augment also holds good for the Laurasian continent. But the

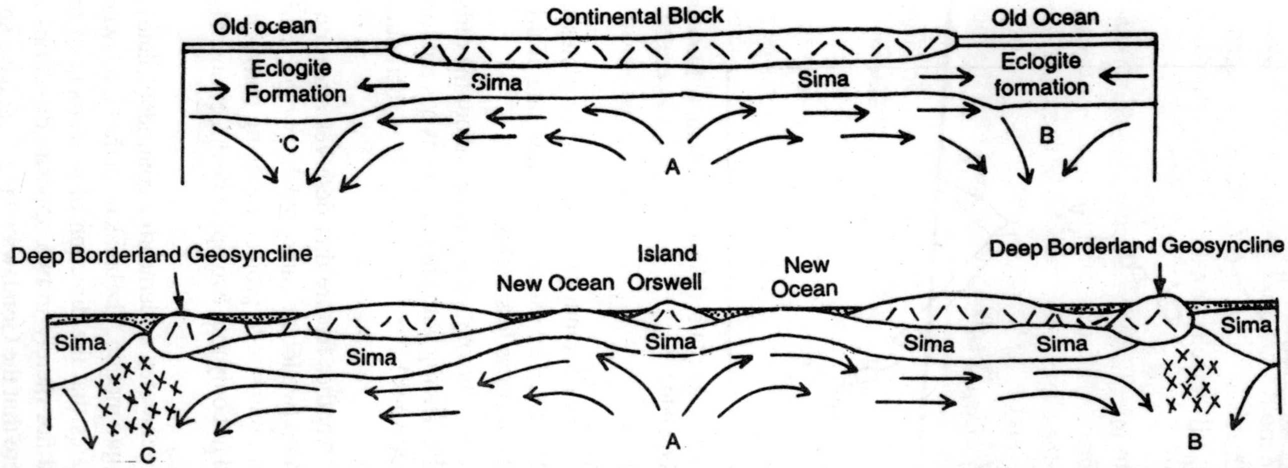

Fig. 19. (After Fig. 29.1 of Arthur Holmes, 1978). Directions of convection currents. Note that the currents break up the original continent into two parts and produce two new continents. Also note that the crustal layer under the continent is considerably thick, while that under the floor of the ocean is very thin.

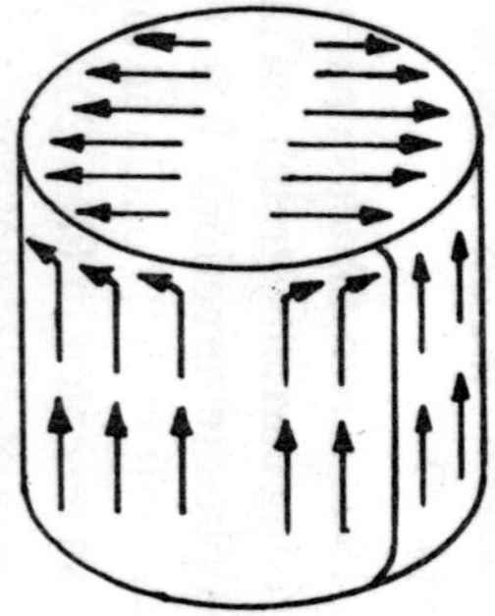

Fig. 20A. Projection of convection currents as per Arthur Holmes. Note that the currents move out in two opposite directions, on reaching the bottom part of the crust (Continent).

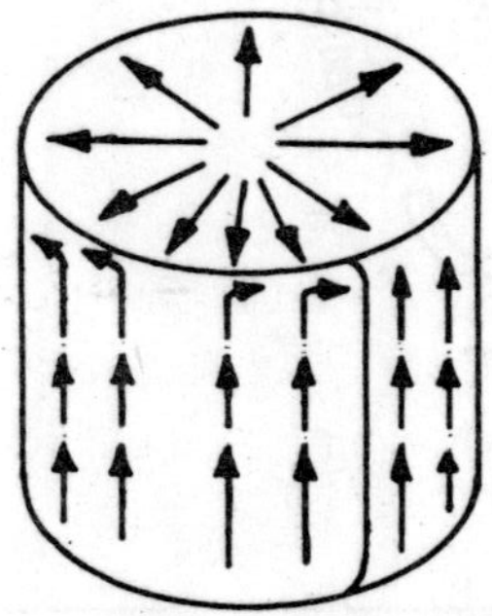

Fig. 20B. Projection of convection currents in a (realistic) Radial Fashion. Note that the currents radiate out on reaching the bottom part of crust (Continent). This is a realistic approach.

feature that has not been understood is, as to what controls the direction of movement of the convection currents! During the episode of the formation of the Deccan Traps of India, magma welled up through the multitudinous conduits. Was the mechanism of convection currents operative during that episode, because the molten material (magma) rose up along the conduits (dykes) into the crustal rocks? All the elaboration indicates that the mechanism of movement of the continents or the plates, still remains a matter of conjecture or a speculation.

Data available through reasearches helps us to infer that the palaeotopographical features during the Permo-Carboniferous period, composed of two continents and one ocean. But the sites of location of the two continents and the ocean, prior to the Palaeozoic period, is still a matter of debate. How were these two continents formed? Could they be of a primordial origin, when the Earth started to cool from its molten state? Considering the fact that the igneous rocks constitute 95% of the continents and the floor of the oceans, the process of cooling of the molten crust, might have resulted in the formation of two large sized landmasses. The possibility of formation of only one huge, gigantic sized landmass also can not be ruled out. The point to be noted is that the postmortem of the topographical features (continents, oceans, mountains etc.) is possible only upto stage of existence of two continents during the Permo-Carboniferous period. But certainly land or landmasses must have existed earlier to that period. But certainly land or landmasses must have existed earlier to that period. What was the "size, shape, number, and location of the landmass or landmasses prior to the Permo-Carboniferous period". Was plate tectonics in operation when the crust of the Earth was being built up during the Archaean period? A schematic evolution of the continents and oceans is presented in Table 8.

Table 8. Scheme of evolution of continents and oceans

CONTINENTS
↓
PANGAEA
(Only one landmass splits into two landmasses)

GONDWANALAND	LAURASIANLAND
↓	↓
(South America, Africa, Antarctica, Australia, India)	(North America, Greenland, Europe, Arctic region)

OCEANS
↓
PANTHALASA
(Only one ocean splits into)

PACIFIC	ATLANTIC	INDIAN

Whether there was only one or more number of continents in the past, the distinction is made on the level of the oceanic water. Atleast the landmasses could be separated from one another, but not the oceans. Distinction of Pacific, Atlantic and the Indian oceans is arbitrary, because there is no barrier in between them at all! Therefore, in reality there is only one ocean and its name is *Hydrosphere*. Real aspect to be investigated is the process of building up of the continental landmasses, and whether these have changed places. Park (op. cit., p. 12) in this context opines, "Despite explosion of research into plate tectonics that has taken place in recent years, we seem to be very little closer to an understanding of the fundamental mechanism that drives plate motion. The ultimate source of energy for tectonic process is heat. The heterogeneous distribution of the heat flow leaving the Earth is converted into density imbalances which in turn provide gravitational energy". Why the gravitational energy thus released, does not induce the plate (continent) to move downwards, rather than in a horizontal direction? *All said and done, the concept of "Plate Tectonics and Sea Floor spreading" is not all proof. It has many loopholes in it.*

STATUS OF PLATE TECTONICS IN THE 21ST CENTURY

From the foregoing account of Plate Tectonics, it is quite apparent that there are numerous loopholes that need rethinking. The 21st century has just commenced, and already the hypothesis is receiving a setback, because, almost every earthquake is being deemed to be caused by drifting of plates. Lamb (2002) writes (pp. 130-131), "In the late 1960s, with the advent of the theory of plate tectonics, geophysicists finally seemed to have made sense of earthquakes. The surface of the Earth is covered in a mosaic of rigid plates, and earthquakes occur in narrow zones where the plates rub against each other. But in 1980s, there were second thoughts. The plate theory worked well in oceans, but across the continental landmasses, earthquakes seem to occur all over the place. Gradually a new view emerged that wasn't plate tectonics at all; in fact, large tracts of the continents seemed to behave more like a fluid. Jackson believes that we must go back to trying to understand the motions of small platelike blocks in the crust - bad news for those studying active fault lines, because as yet, there is no simple way of predicting how these blocks might behave. It now seems that all earthquakes in the continents occur only in the crust; not in the mantle. A more fundamental problem is that geologists have surprisingly little to go on when it comes to understanding the mantle, and there is an embarassingly large gap between laboratory experiments and the behaviour of rocks under real geological conditions. Looking below the crust is not easy, but the rewards in terms of understanding our planet will be worth the effort". Lamb (op. cit.) has based his article "is it all in the Crust" by considering the findings described by Jackson, J.A. (2002), G.S.A. Today, Vol. 12; McKenzie, D.P. and Fairhead, D.J. (1997), Jour. Geophysical Research, Vol. 102.

Plate Tectonics appears to be losing credibility because almost every geological event, be it generation of magma, origin of mountains or oceanic ridges, fault or earthquakes, is being attributed to it. Too much is too bad. Plate Tectonics appears to become a topic of serious debate in the 21st century. Observation of Lamb (op. cit.) that "earthquakes on land are only 'skin deep'. This has prompted a rethink of what gives the tectonic plates their strength", gives a shock to the viability of the process of Plate Tectonics.

CHAPTER 7

HOW, WHY, WHEN AND WHERE OF EARTHQUAKES

INTRODUCTION

The earthquakes are looked at with awe and fear, because of the destruction caused by them on the land, to the buildings, dams and other structures built by man. Is this geological event only destructive or is constructive one, in the context of the building up of the Earth? The point is, why do the earthquakes take place at all, if the process of "construction of the Earth" was completed when it came into being, in the Universe? Even after a period of 4.6 b.y., the Earth appears to be unstable in its interior parts. The geological events enact in pairs (Gokhale, 1996, p. 165), like (i) Plate movement and earthquakes, (ii) Igneous activity and plate movement, (iii) Igneous activity and the earthquakes, and so on. However in each pair which event is first? It is, therefore, necessary to understand the origin and the salient features of the earthquakes.

ABOUT SOME TERMS

In Greek language, the terms *Seio* and *Seismos* mean to *Shake* and *Earthquake*, respectively. Seismology, therefore, gets defined as the "study or discourse of earthquakes". *Seismograph* is an instrument that records the earthquake shocks. *Seismoscope* is an instrument that indicates the occurrence of the earthquakes, but it does not record it. *Seismogram* is the actual record of the earthquake, like the one shown in Fig. 21A, B.

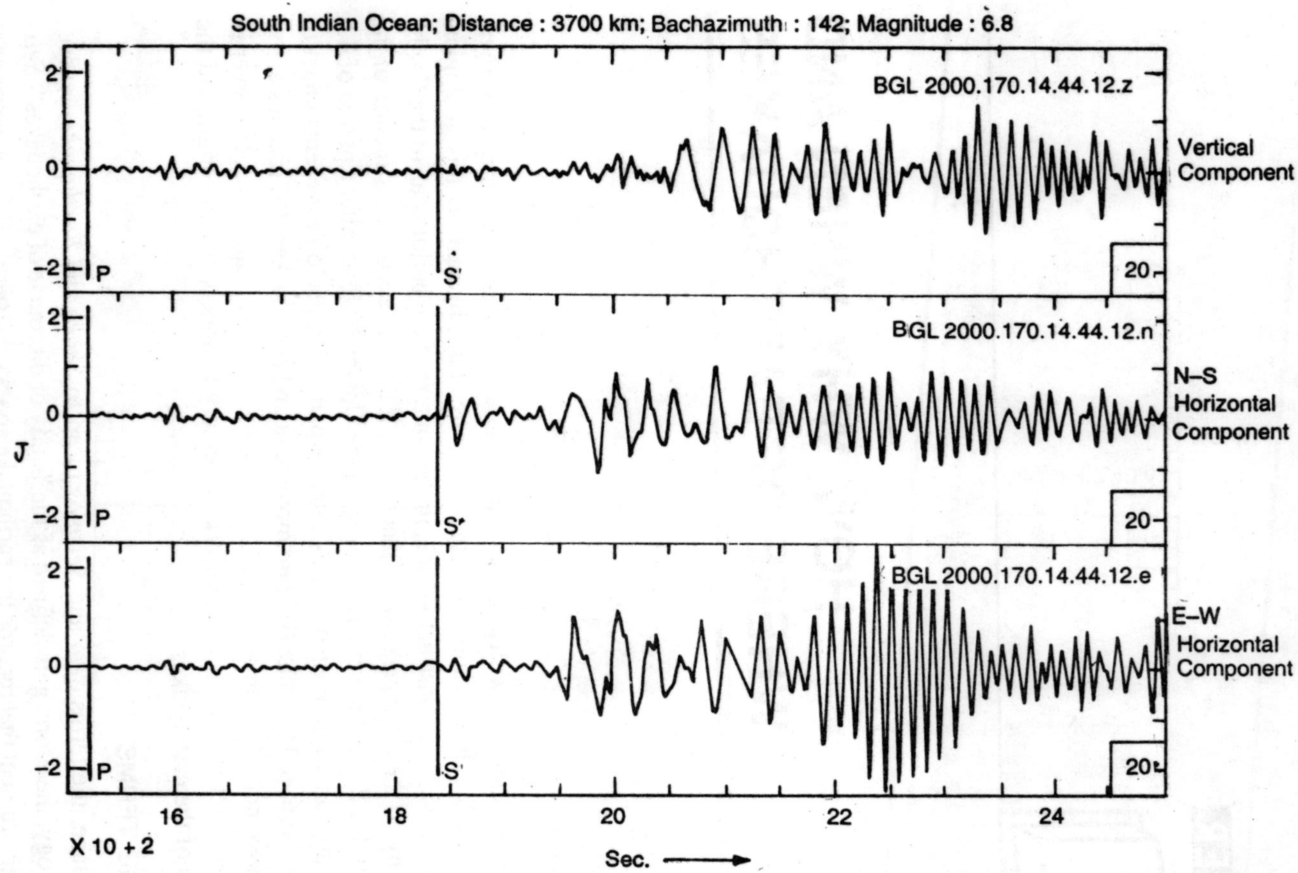

Fig. 21A. Seismograms of earthquakes of magnitudes. (A) 6.8, South Indian Ocean, distance 3700 km, year 2000, (B) 4.5, Bhuj District, Gujarat, India, Lat. 23.68, Long. 70.2, 9th February, 2001 (Courtesy Dr. D. Muralidharan, Scientist, N.G.R.I., Hyderabad, A.P., India).

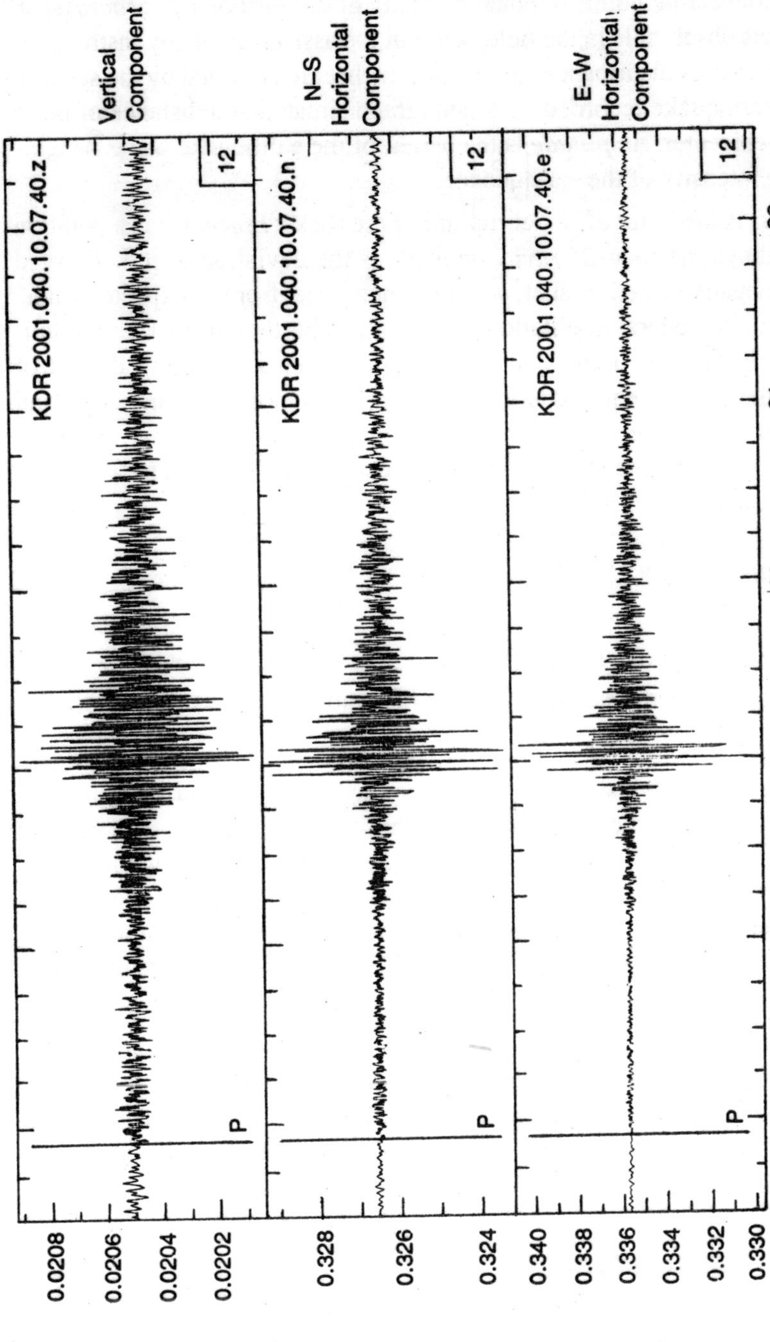
Vertical Component
KDR 2001.040.10.07.40.z
N–S Horizontal Component
KDR 2001.040.10.07.40.n
E–W Horizontal Component
KDR 2001.040.10.07.40.e
12
P
0.0208
0.0206
0.0204
0.0202
0.328
0.326
0.324
0.340
0.338
0.336
0.334
0.332
0.330
17
18
19
20
21
22
X 10 + 2

Fig. 21B.

Seismometer is an instrument whose physical constants are known sufficiently for calibration, so that the actual ground motion could be calculated from the seismogram. *Isoseismal* is a line drawn by connecting points of equal intensity of the earthquake. *Macroseism* denotes large scale effects observable in the field, without the assistance of any instrument. *Microseism* denotes continuous disturbances in the ground that are recorded by the seismograph. *Teleseism* is an earthquake recorded by seismograph situated at a distance of more than 1000 km from the epicenter. *Amplitude* is the height of the earthquake wave. *Magnitude* is a measure of the intensity of the earthquake.

Hypocentre or *Focus* is the point of initial rupture of the rocks. *Epicentre* is a point on the surface, vertically above the focus. *Gravity anomaly* is the deviation of the observed acceleration of gravity (negative when smaller, positive when large) from an expected value, which is calculated from the general gravitational field of the Earth, considering the latitude and the elevation. ERG is the unit of measuring energy released during the act of earthquake. *One erg* is the energy required to make "one gram to move one cm per second faster than it moved during the preceeding second". *Origin time* is the instant when the event of earthquake takes place at the focus. The time is reported as Greenwich Civil Time (GCT), 20, 31, 53, equivalent to $20^h\ 31^m\ 53^s$.

RECORDING OF EARTHQUAKES

It consists of measuring :

(a) the amount of movement of ground, and
(b) intensity of damage caused to surficial features like buildings, dams, railway tracks, and so on.

The movement of ground is measured by a seismograph. The ground motion takes place in three directions; one in vertical, and two in the horizontal directions. The seismological stations are equipped with two horizontal seismographs mounted at right angles to one another; one records N-S, the other records E-W movements. There is one vertical seismograph. These seismographs give a complete record of movements in the three directions. In the Simple Seismograph, pendulum is used. In Fig. 22, the principle of Simple Seismograph is shown.

In Simple Seismograph, the principle of "horizontal bracket pendulum" is adopted. There are three varieties : (a) horizontal pendulum, (b) inverted pendulum, and (c) small torsion pendulum. The weight of the pendulum is of the order of 17 to 21 tonnes. Besides the Simple Seismograph, there are Electromagnetic Seismographs. The seismograph records earthquakes taking place anywhere and at any depth in the crust of the Earth.

CLASSIFICATION OF EARTHQUAKES

The seismographs record a variety of disturbances, the causes for which are different. Richter (1969, p. 151) has classified the "earth disturbances" into several categories, some of which are presented in Table 9.

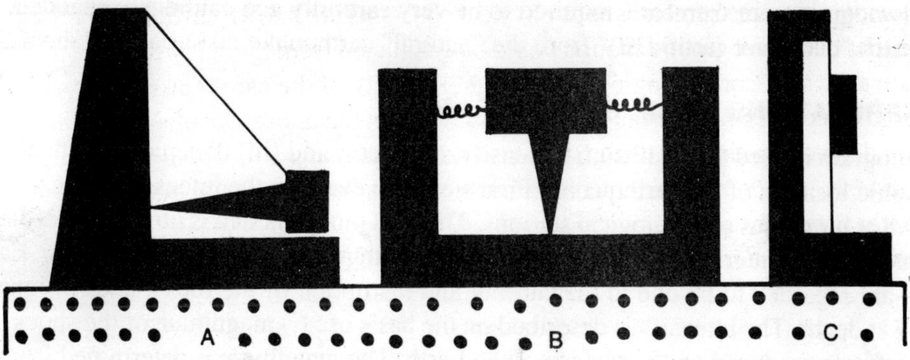

Fig. 22. (After Fig. 153 of C.F. Richter, 1969). Principle of Seismograph using pendulum. A. Horizontal pendulum, B. Inverted pendulum, C. Torsion pendulum. ••• Basement on which pendulums are erected.

Scrutiny of Table 9 reveals that the seismogram record is a mixture of "Artificial and Natural" disturbances. A disturbance due to an artificial cause might be mistaken for a low intensity natural earthquake, because of the similarity between their seismograms. A strong disturbance due to an artificial cause, might as well be mistaken for a natural earthquake.

Table 9. Classification of earth disturbances (after Table 12.1, p. 151 of Richter, in part)

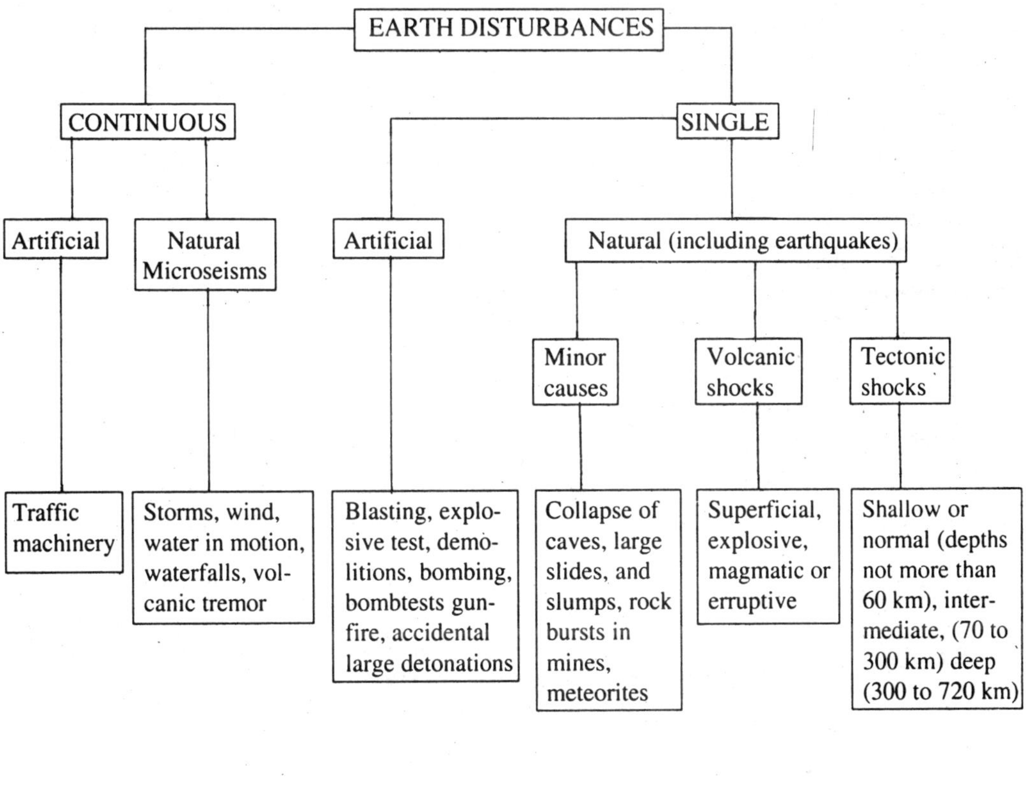

The seismograms, are therefore, required to be very carefully and cautiously decoded, and distinguish "deceptive (artificial)" from the "natural" earthquake disturbances (shocks).

INTERPRETATIONS FROM SEISMOGRAMS

A seismogram is used to establish (i) intensity, (ii) focus, and (iii) the epicenter (that is the geographic location of the earthquake). First step is to evaluate the intensity of the tremors recorded at numerous seismological stations. The data thus obtained is utilised to locate the focus and the epicenter. Derivation of the cause of earthquake is not always possible. Earthquakes are regarded to be due to (a) faulting and disruption of the rocks, and (b) igneous activity at depth. The intensity is described on the basis of : (i) magnitude of the shock, and (ii) the effects produced on the surface of the Earth. The magnitude is determined from the seismogram, while the intensity is judged by the severity of the effects produced by the shocks. The data derived from the seismogram is then utilised to determine the focus and the epicenter, of the earthquake.

ABOUT SCALES USED

The earthquake shocks or the tremors, set the ground in motion. Greater the intensity, greater will be the ground motion. Effects therefore will indicate the intensity of the tremor. The effects are classified into : (i) Primary and (ii) Secondary. The Secondary category is further split into (a) Permanent, and (b) Transient. The primary effects are caused by faulting and volcanic actions. The secondary effects are due to the passage of the elastic waves generated by the primary process. The seismogram thus registers the primary as well as the secondary shocks. Whereas the primary effects are permanent, some of the secondary effects are felt only during the act of the earthquake. These latter effects are called as "Transient". The primary effects of earthquakes produced by the volcanic action are entirely different from those produced by the tectonic earthquakes. Table 10 sets out the effects of the Tectonic Earthquakes as listed by Richter (op. cit., p. 81).

Table 10. Effects of tectonic earthquakes (after Table 7.1, in parts, of Richter, p. 81)

Effects on	*Primary*	*Secondary-permanent*	*Secondary-transient*
Terraine	Regional warping, fissures, elevation or depression of coasts, changes in coastline	Landslide, secondary fissures, sand crater	Perceptible shaking
Water (ground and surface)	Damning waterfalls, change in wells, springs	—	Changes in well leaves, seiches, tsunamis, earthquake fountains, seaquakes
Works of construction	Destruction or damage to buildings, pipe lines, railways, roads	Damage to buildings, windows, plaster	Creaking of frames, swaying of bridges and tall structures
Loose objects	—	Displacement, overturning	Rocking, swinging, shaking, rattling
Miscellaneous	—	Clocks stop, cable breaks, glaciers affected	Sleepers awakaned, animals disturbed, bells ring, audible sound, flashes of light

It is necessary to evaluate the intensity of the earthquakes by considering the distur bances and the destruction caused, in the form of a scale. Such an evaluation does not need any instrument, but consideration of the severity of the effects of the earthquake shocks Rossi-Forel (1883) have set up a scale based on effects upon constructions, and natural objects. Enormous range of intensity is lumped together at the highest level X. The scale is given in Table 11.

Table 11. Rossi-Forel scale (as given by Richter, p. 650)

Grade	*Intensity*	*Effects*
I.	Microseismic shock	Record by instrument
II.	Extremely feeble shock	Felt by a small number of persons at rest
III.	Very feeble shock	Felt by several persons at least. Direction and duration is appreciable
IV.	Feeble shock	Felt by persons in motion
V.	Shock of moderate intensity	Generally felt by everyone
VI.	Fairly strong shock	General awakening of those asleep. Visible agitation of trees and shrubs. Startled persons leaving dwellings
VII.	Strong shock	Overthrow of movable objects. General panic. No damage to buildings
VIII.	Very strong shock	Fall of chimneys, cracks in the walls of buildings
IX.	Extremely strong shock	Partial or total destruction of some buildings
X.	Shock of extreme intensity	Great disaster, ruins, fissures in ground, rock falls from mountains

Mercalli (1931) has set up 12 categories of intensities for the earthquakes, on the basis of effects. The categories given by him are presented in Table 12.

It is felt necessary to elaborate some of the effects that are the indicators of the act of the earthquakes. However, these are produced within the span of the earthquake shock. These are described below.

(a) *Oscillation of water level in wells* : A passing seismic wave compresses the acquifer. It squeezes additional water into the well, raising its level. The opposite half of the wave, dilates the acquifer and draws water out of the well causing depression in level (Richter, p. 106).

(b) *Earthquake fountains* : When there is plenty of groundwater, a strong earthquake often produces fountains, "geysers", which play during the strong shaking and for sometime afterward (Richter, p. 106).

(c) *Seiches* : These are "standing waves" set up on the surface of an enclosed body of water, such as lake, pond or tank (Richter, p. 109).

(d) *Seaquakes* : Shaking caused by earthquakes, and it is felt on board a vessel at sea. Tremendous grating motion of ship, as if grazing over a coral. No ripples on the

Table 12. Mercalli scale (1931) (as given by Richter, pp. 137-138)

Category	*Effects*
I.	Not felt
II.	Felt by persons at rest on upper floors
III.	Felt indoors. Hanging objects swing. Vibrations like passing of light trucks
IV.	Vibrations like passing of heavy trucks. Crockery clashes. Windows rattle
V.	Felt outdoors. Direction estimated. Sleepers awakened. Doors swing and close
VI.	Felt by all. Many frightened and run outdoors. Persons walk unsteadily. Windows broken. Books fall off from shelves
VII.	Difficult to stand. Noticed by drivers of motor cars. Fall of plaster, loose bricks, stones and tiles. Damage to masonry
VIII.	Steering of motor care affected. Fall of monuments, towers, elevated tanks. Branches broken from trees. Changes in flow or temperature of springs and wells. Cracks in wet ground and on steep slopes
IX.	General panic. Masonry heavily damaged. Underground piper broken. conspicuous cracks in ground. Sand, mud ejected. Sand craters
X.	Most masonry structures and foundations destroyed. Large landslides. Water thrown on banks of rivers and canals
XI.	Rails greatly bent. Underground pipe lines completely out of service
XII.	Total damage. Large rock masses destroyed. Objects thrown into the air

surface of sea, but thunderous sound emerges from the ship's bottom. A feeling as though the ship got anchored in deep sea water, the depth being of the order of 40 fathoms without touching bottom. The vessel shook violently, a rumbling, grating sensation was experienced. Masts, funnel and superstructures vibrated and rattled alarmingly, giving the impression that the ship was running a ground on to hard bottom and buckling. The chart shows 1800 to 1900 fathoms water in the vicinity (Richter, p. 118).

(e) *Sound* : It is due to the rattling and shaking of loose objects indoor. The true earthquake sound is heard more clearly out of doors. It is usually a heavy sound, pitched near the lower limit of hearing, and most commonly compared to thunder, gunfire, or heavy traffic at a distance (Richter, p. 128).

QUANTIFICATION OF INTENSITY OF EARTHQUAKES

Earthquake shocks are due to a sharp displacement of the crustal rocks at depth. Depending upon the amount of displacement, the severity of shocks will vary. It is, therefore, necessary to measure the displacement, which is, however, not feasible, because it takes place at great depths. Effects on the surface can not be used to derive the displacement. The earthquake shock sends elastic waves in the crustal rocks, and those are recorded on the seismogram. Nature of the wave (its length and amplitude) is related to the intensity of the shock.

This data forms the base for constructing a scale of intensity of the earthquakes. No doubt there is a definite correlation between the intensity of the shock and the degree of damage caused on the surface. However, there will not be any damage accompanying a severe earthquake, if it takes place in deserts, oceans, or in mountainous tracks, because three will not be any human habitation and structures like buildings, rail-tracks and so on. The location of the epicenter with respect to the human habitation also has an effect upon the degree of damage. In this regard Richter writes (op. cit., p. 338). "In the minds, Bakersfield, California, the major earthquake (magnitude 7.7) of July 1952, much less important than comparatively minor aftershock (magnitude 5.8) on August 1952, which originated closer to the city and was accompanied by much more local damage, representing higher local intensity (VIII against VII)". Richter further writes "Rating in terms of intensity was hardly possible; many of the shocks were not reported as felt, and some obviously large ones were reported only as barely perceptible at point distant, from epicenter, which might be in a thinly populated desert or mountain area, or off the coast (pp. 339-340).

Need is thus felt to evolve a yardstick to quantitywise assess the earthquake shocks. The efforts of several seismologists (Gutenberg, C.E., Richter, C.F., Bath, M., Hayes, R.C., diFillipo, D., and Marcelli, L., Zatopak, A., Vanek, J., Tsuboi, C., Wadati, K., kawasumi, H., Karnik, V., Honda, H.) led to the scalar concept. The terms "degree and intensity" of earthquakes were replaced by the term "magnitude". The seismogram consists of recording of waves emanating from the focus of the shock. Amplitude of an earthquake not only depends upon the magnitude and distance, but also upon the physical conditions along the path, and the ground conditions at the recording station, as well as with the characteristics of the seismographs used.

The amplitudes of the waves are measured and are plotted on a logarithmic scale, because the measurements range from 0.1 mm to 10-12 cm. The value of the magnitude is calculated by solving equation which is given below.

$$M = \log A \text{ minus } \log A_0$$

where A = Recorded trace of amplitude for a given earthquake, at a given distance as written by the standard type instrument.

A_0 = is the trace of amplitude for a particular earthquake selected as standard, and

M = Magnitude

The magnitude is thus a number characteristic of the earthquake and is independent of the location of the recording station. Three arbitrary choices enter into the definition of M. These are :

1. Use of a particular type seismometer with specified constants,
2. Use of ordinary logarithms to the base 10, and
3. Selection of the standard shock whose amplitudes are represented by A_0. The standard shock is also called as zero shock.

The magnitude scale for the earthquakes ranges from 3.5 (feeblest) to 8.1. Maximum magnitude so far recorded was in Chile, 1960, it is being 8.9, when three major shocks were felt. Though numerous seismologists have measured the magnitude of the earthquakes of different bases, the one proposed by Richter is the most popular one, and it is in vogue. The earthquakes are, therefore, described by reporting the value on the Richter scale. However, it is necessary to remember that an earthquake of lower magnitude can produce considerable damage, when it affects a populated region. The inference drawn is that it was a major earthquake, which is incorrect. Likewise, an earthquake of higher magnitude might produce feeble destruction, if the epicenter is located far away from populated area. The inference drawn of a minor earthquake also will not be correct. The intensity of earthquake should, therefore, be assigned from the value on the Richter scale and not merely from the effects of the shocks.

Holmes (1978) gives a correlation between the effects of the earthquakes on the Mercalli scale, and the values on the Richter scale. It is given in Table 13.

Table 13. Correlation between Mercalli and Richter scales (after Holmes, 1978, p. 575)

Grade	*Mercalli scale Effect*	*Richter scale Magnitude*
I.	Instrumental	Upto 3.5
II.	Feeble	3.5 to 4.2
III.	Slight	
IV.	Moderate	4.3 to 4.8
V.	Rather strong	
VI.	Strong	4.9 to 5.4
VII.	Very strong	5.5 to 6.1
VIII.	Destructive	6.2 to 6.9
IX.	Ruinous	
X.	Disastrous	7.0 to 7.3
XI.	Very disastrous	7.4 to 8.1
XII.	Catastrophic	> 8.1

MAGNITUDE AND ENERGY

The intensity of the earthquakes is also evaluated from the energy released. At the focus, the act of disruption of the rocks, releases energy. A part of the energy is expended in the transmission of the elastic waves in the crustal rocks around the focus, and a part is dissipated as heat. The original potential energy of stress, stored in the rocks, performs mechanical work, like lifting the crustal blocks against gravity, crushing material in the fault zone. Energy released is proportional to the square of the amplitude. Gutenberg and Richter (1955, as noted by Richter, op. cit., 1969 on page 365) have putforth an equation for calculating the energy released. The equation is given below.

$$\log E = 5.8 + 2.4\,m$$

where M is the magnitude (like 3.5, 4.0, 7.5 and so on), and
E stands for energy

Energy released is measured in ergs. Some examples of values of energy released during earthquakes are given below.

1. California earthquake of 1906; M = 7.8, energy released 1.75×10^{24} ergs,
2. Japan, 1929, M = 7.1, energy released 3.1×10^{20} ergs,
3. "Nominal" atomic bomb of Hiroshima type is 8×10^{20} ergs. Energy released by largest earthquake is not much over 10^{25} ergs, roughly equivalent to 12,000 of the nominal bombs (Richter, p. 366).
4. Earthquake of 8.5 magnitude occurred only three times in the 20th century, its energy corresponds to $10^{26.4}$ ergs (Holmes, 1966, p. 902), and
5. Average annual release of energy from all earthquakes ranges between 10^{25}-10^{27} ergs.

MAGNITUDE AND FOCUS

Frequency of shocks generally decreases with increasing depth. From the records of 5605 shocks in Italy, Oldham (Holmes, 1966, p. 904) found that 90 per cent of the earthquakes originated at depths of less than 8 km, nearly 9 per cent at depths between 8 and 30 km, and the rest at greater depths. In most areas, there is a fairly definite lowest horizon, the range is of 600-700 km. No shocks are known below 700 km. Depths of 60, 70 to 300, and greater than 300 km, are classified as shallow, intermediate, and deep focus, respectively. Deepest known shocks are given below :

(i) Flores sea, depth 720 km of magnitude 7.1,
(ii) Spanish shock, depth 640 km of magnitude 7.1, and
(iii) Earthquake of 8.6 magnitude occur at shallow, and intermediate depths (Richter, op. cit. Tables 22-5, 22-6, 22-7, pp. 350, 351, 352).

It is thus apparent that the magnitude of the earthquakes is not dependent on the depth of the focus. Cause for the absence of shocks at depths greater than 700 km, is that at those depths, the rocks are nearly in a plastic state, and therefore, they cannot withstand the stresses.

CLASSIFICATION OF SHOCKS

A seismograph records numerous shocks that are originated by a variety of causes. The tremors belong to two main sources : (a) continuous disturbances, and (b) single disturbance. Seismogram produced by the first cause is uniform, because the source of disturbances (traffic, machinery, storms, wind, frost, streams, waterfalls, volcanic erruptions) sends out shocks continuously over a considerable period of time. Seismogram produced by the second cause is sharp, short and with considerable variations, because the source (blasting, explosive tests, bombing and bomb tests, gunfire, rock bursts in mines, meteorites, tectonic shocks) send out shocks over a very short time of a few seconds, a minute or so.

The shocks are categorized into : (a) fore, (b) main, and (c) after. The foreshocks seldom afford any oppurtunity for warning or prediction of major earthquake, because these are mild. Shocks produced by "non-earthquakes" category (traffic, storms, machinery etc.) have a resemblance to the aftershocks of true earthquakes. Besides the "main, and after" shocks, there is one more variety. It is called as "Earthquake Swarms". These are common in the volcanic regions, and occur before and during the volcanic erruptions. These tremors are more or less continuous, and are recorded near a volcano. These are numerous, small individual shocks.

NATURE OF GROUND MOTION

The direction of ground motion will depend upon the cause of the earthquake. Earthquakes are caused by the tectonic process (faulting) and by the volcanic erruptions. In order to ascend upwards, the lava builds up pressure in the crust, and when the pressure crosses the limit the crust could bear, violent explosion takes place, and the lava errupts on the surface of the Earth. The act of explosion shakes the Earth. The direction of movement of lava being upwards, the ground also is shaken in that direction. Therefore, the ground motion will be in a vertical direction. The elastic waves are, therefore, sent out in the crust, more in the vertical direction. In the tectonic earthquakes, the crustal rocks move on the fault plane, either in a horizontal, or in a vertical direction. The elastic waves produced by such a movement will be prominently guided by the direction of the movement along the fault plane. Thus the direction of ground motion depends upon the cause of the earthquake.

HOW, WHY, WHERE AND WHEN OF EARTHQUAKES

Man anxiously strives to know in advance, "when and where" the earthquakes might take place. The solution to this lies in understanding the "why and how" of that geological event. Earthquakes have taken place since Earth was born, the activity is continuing at present, and will be recurring in the future too. It is one of the geological episodes that enact in the Earth. The effects of this episode being destructive, it is worthwhile to ponder over the possibility of predicting the time and the place of the earthquakes, in advance, so that the destruction could be minimised. Let us discuss the "Why, How, Where and When" of the earthquakes. How do the earthquakes originate? The cause is determined from the effects produced by the shocks. Majority of the earthquakes are ascribed to the disrupting and rupturing action taking place during the faulting of the crustal rocks. This tectonic movement triggers off earthquake shocks. This in turn raises the question of "what initiates faulting". To bring about disruption in the crust, generation of "strong motive force" at depth is essential. Such is the background of the earthquakes originating from faults. One of the effects of the volcanic erruptions is sending tremors in the region surrounding the volcanic center. This episode forms the second cause of the earthquakes. The modus operandi stems from the generation of magma at depth, which tries to force its way upwards through the crustal rocks. In this act, the magma tears apart rocks at some place and culminates into an explosive volcanic erruption.

The earthquakes thus originate due to (a) tectonic process, and (b) volcanic erruptions. This leads to the possibility of existence of a "latent remote control" operative in causing disruption (faulting) of the crustal rocks, and formation of volcanic rocks as an outcome of the volcanic erruptions. As far as the Earth is concerned, these two events are of a "constructive nature", because these built up and do not destroy the crust of the Earth. The Earth is shaken not with the intention of destroying human habitation, or the buildings, dams and other structures. Considered in this apparel, the earthquake events should be regarded as "constructive episodes".

The "Why and How" of the earthquake is closely inter-related to one another. Earthquakes are the manifestations of faulting of rocks, and the volcanic erruptions. Faulting requires generation of stresses in the deeper parts, while the volcanic erruptions need generation of magma at depth. This leads us to discuss the modus operandi of creating stresses and magma, at depth. What is the compulsion on the part of the Earth of generating stresses and magma? Does it reflect instability of the Earth at depth?

Let us now turn to consider the "When" of the earthquakes. Are these controlled by the parameter of time? How much frequently do the earthquakes occur? Oldest recorded earthquake dates back to 373 B.C. Frequency of the earthquakes has been set out in Table 14, the contents of which have been taken from Table 22-9 of Richter (op. cit., p. 356). It sets out number of large shocks only.

Table 14. Frequency of earthquakes

Depth	Amplitude	Year		
		1897-1903	1904-1917	1918-1955
Shallow	8.6 and above	6	7	9
	7.9 - 8.5	32±	25	66
	7.0 - 7.8	–	–	570
Intermediate	8.6 and above	–	3	1
	7.9 - 8.5	–	12	8
	7.0 - 7.8	–	61	244
Deep	7.9 - 8.5	–	4	4
	7.0 - 7.8	–	18	66

In Table 14, the statistics is given only for the large shocks. It indicates that over a span of 58 years (1897 to 1955) there were 715 Shallow, 299 Intermediate, and 92 Deep focus shocks. Percentage-wise it works out as, 64.6, 27.0 and 8.3, respectively for Shallow, Intermediate and Deep focus shocks. Magnitude-wise, out of a total of 1106 shocks, 929 belong to 7.0-7.8 category, while only 177 belong to 7.9-8.5 and above, category. Percentagewise it works out as 84 for 7.0-7.8, and 16 for the other category of magnitude. Scrutiny of Table 14 further reveals that there is a rapid increase in the frequency of shocks from 168 during 1897-1917 to as high as 938 during 1918-1955.

Oldham has noted that 90 per cent of shocks occur at shallow depths. Table 14 supports it, as 64 per cent shocks belong to shallow depths. Malik, J.N. et al. (1999) have reported that out of a total of 89 earthquake shocks that had occurred in Kachch Peninsula, Western India, during 1668-1996, as many as 87 are of shallow focus with a magnitude of 2.6 to 6.0. The remaining two shocks too, are of shallow depth but have a magnitude of 7.6 and 7.8. The observation of Malik et al. (op. cit.) too, indicates that the percentage of shallow depth shocks is as high as 97. Lamb (op. cit.) holds the view that "New observations suggest that earthquakes on land are only 'skin deep', confined to the Earth's outermost layer of crust. This has prompted a rethink of what gives the tectonic plates their strength?" *The statistics thus reveals that the earthquakes are not related to any definite period, nor there is any periodicity of occurrence. Uncertainty, therefore, prevails regarding when the earthquake will take place.*

The last facet of the earthquakes is to consider the place where these might occur. Plotting of centers of the earthquakes has given rise to three linear belts : (1) the entire coastline of the Pacific Ocean, (2) E-W trending Mediterranean belt that extents upto Indonesia, and (3) Mid-Oceanic ridges, located in the Atlantic ocean. It is pertinent to observe that these belts also mark the sites of volcanic erruptions, as well as the margins of the continental and the oceanic plates. The feature to be noted is that the earthquakes, however, also occur outside the said belts. The Peninsular India was earlier categorised as a "Shield Area" where the earthquakes will not be taking place. But the instance of the occurrence of the earthquake in the Koyana region of Maharashtra, India, shook the concept of "Shield Area". Occurrence of epicenters in Tamil Nadu, Kerala states of India, which are situated considerably south of the earthquake prone Himalayan mountains, indicates that the earthquakes could take place anywhere. There is thus no control of geographical location upon the sites of epicenter of the earthquakes.

From the foregoing account of the "How, Why, When and Where, it is obvious that the earthquakes can not be predicted with certainty. The hypothesis of Plate Tectonics attributes the earthquakes to be due to the movement of the Continental and the oceanic plates. But to set the plates in motion, driving force is required to be created at depth. Therefore, the episodes of Plate Tectonics, earthquakes, and the volcanic erruptions, are initiated in the deeper parts of the Earth. The episodes are genetically related to one another, and these build up the crust of the Earth. Even after a period of 4.5 b.y., the building activity of the Earth is still continuing. The Indian mythological interpretation of the cause of the earthquakes to be due to the jerk given to the Earth by a giant sized serpant named *Shesh*, is misunderstood. The real meaning is that the Earth is shaken by the forces acting from within, and not from the bottom part of the Earth. A common man finds it very difficult to realise that the Earth is not only not resting on anything, but is spherical in shape and is suspended in empty space. The serpent infact means, an internal force causing earthquakes. For the consumption of the common man, the Earth is projected to be resting on the hood of the serpent *Shesh*.

There is a growing tendency to ascribe the earthquakes to Plate Tectonics, whenever no other cause is apparently visible. Geoscientists, however, have to ponder over extending such an explanation. The Alpine-Himalayan Mountain Ranges are regarded to be the outcome of the movement of the Indian and the Eurasian plates towards one another, the event having taken place, way back in the Eocene period. The two plates, therefore, are in juxtaposition since that period, leaving no space in between them. Yet, the earthquakes that had occurred in the Gujarat region of India during 2000, are ascribed to the movement of the Indian Plate in a northerly direction, notwithstanding any evidence of the movement of the said plate. The cause of the earthquakes of the Gujarat region, India, could as well be different one. At this juncture, the earthquake in the Etna region, Sicily, on 31.10.2002, is thought provoking. The shock measured 5.4 on the Richter scale, and occurred in the vicinity of Etna which was accompanied by erruption of volcano of Mt. Etna, the outpouring of lava continuing for 4 successive days. This episode clearly demonstrates that the lava while forcing its way upwards through the crustal rocks, caused the earthquake in the region surrounding it. The inference to be drawn is that the magma generated at depth found an exit-way to the surface, and welled out as lava. It is, therefore, also possible that the igneous activity (generation of magma) creates tension at depth, and the molten material (magma) tries to rise upwards, but it does not always succeed in reaching the surface of the Earth. The hypothesis of Convection Currents of Holmes, is of a similar nature. The upward moving magma could shake the crustal rocks and send earthquake tremors. Thus igneous activity taking place at depth could be the cause of some of the earthquakes, and not necessarily due to the movement of the plates.

OPINION ABOUT WHY AND WHERE OF EARTHQUAKES IN 21ST CENTURY

Whether the earthquakes are due to the movement of plates, is going to be a debatable issue in the 21st century. The question of, which is the first event, Earthquake or Plate Movement, is going to take the center-stage. Lamb (op. cit.) has raised the issue of force that can set the movement of the plate. The opines that the confinement of the earthquakes which are 'skin deep' to the Earth's outermost layer of crust, it is necessary to rethink about what gives the "Tectonic Plates" their strength. Existence of plates and their movement, is going to be a burning topic of scientific research in the 21st century.

CHAPTER 8

SOME FEATURES OF MAGMA, ITS CRYSTALLISATION AND CHEMICAL COMPOSITION OF EARTH

95 per cent of the crust of Earth is composed of igneous rocks, which are mainly made up of silicate minerals. These rocks were primarily in a molten state. All the elements (Si, Na, Fe etc.) are formed from the burning of hydrogen (Fig. 1). Considering the dominance of the silicate minerals, condition appears to have been extremely favourable for the generation of large quantity of silicon (Si) during the process of the formation of the several elements from the burning of hydrogen. The various minerals are formed at different temperatures; some at a very high temperature (Pyroxenes, Olivine, Anorthite, Bytownite etc.), some at low temperature (K-Feldspars, Quartz, Albite, Oligoclase etc.). These minerals are the outcome of cooling and the concomitant process of crystallisation of the magma (Fig. 23).

The quality of the minerals (felspars, augite, hornblende, quartz etc.) and their quantity (content of the individual minerals), depends primarily on the composition of the magma (Acidic or Basic).

MAGMA A MELT, NOT A SOLUTION

The temperature of the magma is generally around 1000°C, and that temperature, all the constituents of the magma, are in the molten state. As the magma cools, minerals commence crystallising as soon as their "temperature of formation" is attained, and not because

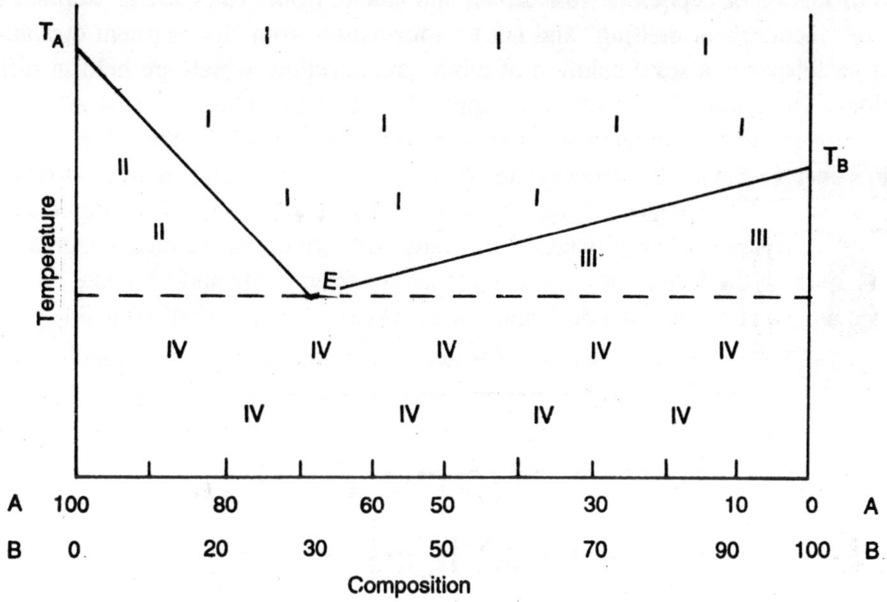

Fig. 23. Crystallisation of binary system (note that the curves indicate the temperature of melting/freezing and not that of saturation of minerals A and B). T_A, T_B : Melting or freezing points of A and B; T_AE : Curve of melting/freezing of A; T_BE : Curve of melting/freezing of B. E : Point of melting/freezing of both A and B. It is designated as the Eutectic point; I : Melt of A and B; II : Crystals of A, melt of B; III : Crystals of B, melt of A; IV : Crystals of A and B.

the "temperature of saturation" of the minerals was reached, by the magma. The term "saturation" thus causes confusion, because the magma is a melt and not a solution of the minerals, in any solvent.

MELTING POINT, NOT A SATURATION POINT

The first formed magma which is designated as the *Primary magma*, no doubt, it is to be attributed to the initial molten stage of the Earth. Subsequently, however, in the past period of 4.6 b.y., igneous rocks have been formed several times, which requires the regeneration of magmas through the process of remelting of the crustal rocks. The "Anatectic Magma" belongs to such a category. The course of crystallisation followed by the Binary and the Ternary magma systems, is traced by plotting the values of "melting points" against the "composition" of the melt (magma) (Fig. 23).

The curves (Fig. 23) thus obtained, are however, described as the "curves of saturation" of the minerals. As already pointed out, the minerals crystallise because the melt (magma) had reached their temperature of "formation". Therefore, the diagrams depicting the course of crystallisation of the minerals from the magma, should name the curves as "curves of formation of crystallisation" and not as "curves of saturation".

The formation of nepheline from albite, and laucite from orthoclase is ascribed to the process of "incongruent melting" and not to a formation from "incongruent dissolution". Plagioclase felspar is a solid solution of albite and anorthite which are held in different proportions. The phrase "solid solution" applied to the plagioclases, conveys that albite and anorthite are not distinguishable as two independent minerals, the same way as the different salts also can not be differentiated from one another, when held in solution. Albite and anorthite are the silicates of Na, Ca, and Al. There are albite-rich (Oligoclase) and anorthite-rich (Bytownite) plagioclasses, and these are formed from a melt (magma). In the diagram depicting the formation of the plagioclasses from albite and anorthite, the curves are named as "Solidus and Liquidus" and not as curves of "Saturation" (Fig. 24).

The "Continuous Reaction Series" of Bowen lends support to the inference that "temperature and not saturation" governs the process of magmatic crystallisation. Thus at high temperature "olivine, bytownite, anorthite", while at lower temperature, "andesine, oligoclase, albite, K-felspars, quartz" are formed from the melt (magma). Thus the minerals of

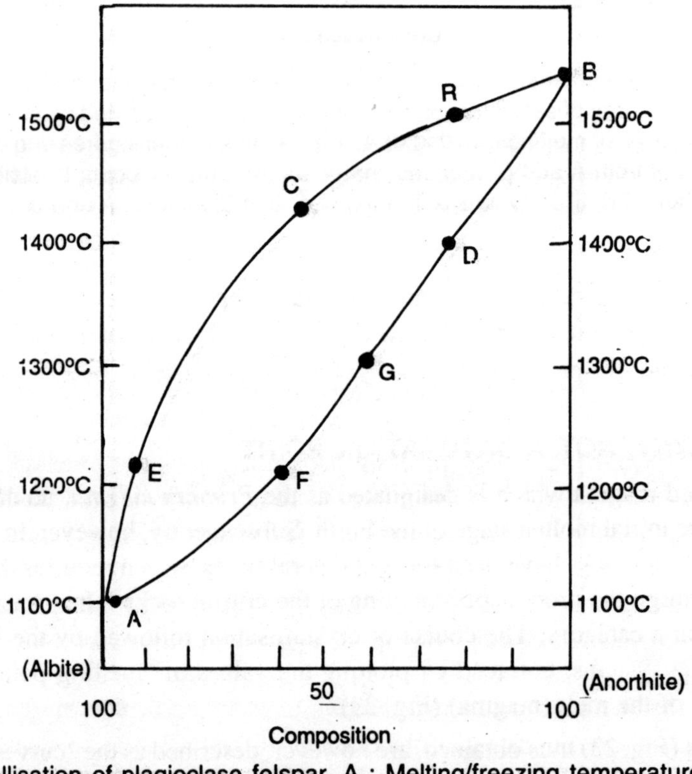

Fig. 24. Crystallisation of plagioclase felspar. A : Melting/freezing temperature of Albite; B : Melting/freezing temperature of anorthite; AECRB : Liquidus curve (indicates composition of melt; AFGDB : Solidus curve (indicates composition of Plagioclase felspar). Note : The curves indicate temperature of melting/freezing and not that of saturation of the minerals Albite and Anorthite. The plagioclase felspar belongs to the category of "Solid Solution".

the Continuous and the Discontinuous Reaction Series, are formed according to the magma reaching the temperature of formation, and not that of "saturation" of the minerals.

If the magma be deemed as a solution and not a melt, then what is the solvent that held the minerals in a dissolved state? The silicate minerals are soluble only in hydrofluoric acid (HF). Large volumes of HF acid, are therefore, required to be generated at depth, in order to dissolve the silicate minerals of the crustal rocks. As this acid does not enter into the composition of the silicate minerals, it becomes imperative to tender an explanation for its disappearance, after completion of the act of crystallisation of the magma (melt). One of the constituents of the magma (melt) is hydrosilicic acid (Orthosilicic, Metasilicic and Polysilicic, corresponding to H_4SiO_4, $H_4SiO_2O_6$, and $H_4Si_3O_8$), besides SiO_2, Na_2O, K_2O and many others. The role of hydrosilicic acid, is however, to build up the silicate structure of the minerals from Al, Na, Ca, Mg, Fe and so on, that are present in the magma (melt), and not forming the "solvent". Thus the "Orthosilicate, Metasilicate and the Polysilicate minerals, examples being the Olivine, Enstatite, and Orthoclase, respectively are formed", besides many others. One of the physical properties of the minerals is their "temperature of melting (M.P.)" which, however, is not designated as the "temperature of solution".

In the background of all the aforesaid features, the term "saturation-curve, and solution" create confusion, because they are misleading in the context of crystallisation of minerals from the magma (melt). It is felt proper to replace them by "melting-point curve, and melt" respectively. The Silica-Saturation Principle of Shand, apparently indicates that the minerals are formed from a solution. In reality, however, the term "saturation" refers to the *degree of silication* acquired by the minerals, like the orthosilicate, the metasilicate etc., which in turn depends upon the quantity of SiO_2 present in the melt (magma). Therefore, the kind of the minerals formed, depends upon the composition of the melt (acidic or basic magma), and their temperature of formation (freezing point), but not because the melt (magma) had reached the temperature of saturation of the minerals.

The "Reaction Principle" of Bowen clearly brings out the fact that the minerals of the magma were in a molten state. The early formed minerals, therefore, could react with the remaining melt, and build up new minerals that are stable at a lower temperature. The term "reaction" explicitly conveys that the minerals are not formed due to reaching the stage of "saturation" in any solvent. The occurrence of recrystallised xenoliths in the granites and basic dykes, reveals the role of a "melt, and not that of a solvent". Gokhale and Koppad (1971) and Gokhale and Puranik (1981) have reported reaction around xenoliths of quartz caught in basic dykes. Gokhale (2001) has reported the occurrence of Orbicular Structure in the gray granitic rocks exposed near Idgah Hill, Saundatti Town, Belgaum District, Karnataka State, India. The structure is due to the reaction around basic xenoliths caught in the granites. The very foundation of the concept of the "Convection Currents" of Holmes (1960) is based on the generation of magma at depth, through the process of remelting of the mantle. Lavas errupted onto the surface of the Earth, are in a molten state, and not in a state of solution.

The process of metamorphism is defined as a response in the heart of solid rocks, to the pronounced changes in temperature, pressure and chemical environment taking place below the shells of weathering and cementation. The pre-existing rocks recrystallise through localised melting of the constituent minerals, and not from a solution. All said and done, metamorphism too indicates that the minerals recrystallise due to generation of temperature of their "formation" and not that of "saturation".

DISPOSITION OF MINERALS IN IGNEOUS BODIES

Let us now turn our attention to the disposition of the minerals in the igneous bodies, like the stock, the boss, the batholith and so on. It is seen that the several minerals (felspar, quartz, diopside, hornblende etc.) are dissiminated throughout the extent of the igneous bodies. Each mineral forms a crystal. A crystal grows around a center. Therefore, there is the necessity of creation of several centers around which the crystals (minerals) could commence their growth. The aspect to be pondered over is, whether the constituent minerals were already in existence in a molten state, or, whether the several elements required for their growth (Na, Ca, Mg, Si, Fe, K, Ti, Al etc.) were transported to the several centers of crystallisation. The latter mechanism involves sharing of the elements, because these are required for the building up of the different minerals. Niggli (quoted by Tyrrell, op. cit., p. 50) has postulated 14 molecules that form the principle constituents of the magma. Reaction between these molecules during the crystallisation of the magma (melt) gives rise to the minerals. Thus the reaction between Kaliophilite and quartz, forms orthoclase, that between Nepheline and quartz, forms albite and so on.

The accessory constituents (ilmenite, rutile, magnetite, zircon and so on) of the igneous rocks are the first minerals to crystallise from the magma, and these are followed by the silicate minerals. The feature to be noted is that though considerable quantity of iron (Fe) is available in the magma (melt), yet only a few grains of magnetite, ilmenite, rutile etc., are formed, while more iron (Fe) is bound up in the later formed ferromagnesian minerals. Does this feature indicate that iron (Fe) was already bound up in the ferromagnesian minerals, which however, were in a molten state?

From the foregoing account of the disposition of the minerals in the igneous bodies, and the other aspects, it is quite evident that the minerals form from a melt, and not from a solution. Magma, therefore, is a melt and not a solution.

CONTRAST IN THE DISTRIBUTION OF IRON IN THE EARTH

Iron attracts attention because

(i) it is the major constituent of the core having a radius of 5200 km (Table 2),
(ii) it is indispensable in the steel industry,
(iii) it causes the magnetic field of the Earth,
(iv) it renders the core denser, thereby resulting in the effective density value of 5.527 (Table 2) of the Earth,
(v) its average content in the crustal rocks that extend to a depth of 150 km, is a mere 6.88% (Table 5), and

(vi) pre-Cambrian period in the Earth's history, simulates an "Iron Age", because almost the entire iron ores, and the iron bearing formations belong to that period.

Iron is one of the elements formed when the Earth came into being some 4.6 b.y. ago. As described in Chapter 2, all the elements composing the entire Earth were formed through the process of burning of hydrogen at a very high temperature (Fig. 1). Though the content of 6.88 per cent of iron ($FeO + Fe_2O_3$, Table 5) apparently appears to be small, it however, accounts for a thickness of 150 km of the lithosphere. Considering the radius of 5200 km of the Core, and iron being its major constituent, the content of iron will superceed even that of SiO_2, when the overall chemical composition of the entire Earth is taken into account. Significance of this feature will be appreciated on considering the fact that in the crustal rocks (150 km thick), SiO_2 dominates all the other elements, $FeO + Fe_2O_3$ taken together occupies 3rd position (Table 5).

Dott et al. (op. cit.) have advocated the process of density differentiation for the disposition of the various elements in the Earth. Accordingly, in the initial stage of accretion, the elements were disseminated throughout the Earth (Fig. 2a). Later, the gaseous mass of the Earth started condensing. The lighter elements moved towards the peripheral parts, while the heavier ones (Fe, Ni) moved towards the central part of the Earth (Fig. 2b). Williamson and Adams (1923), and Goldschmidt (1922) have extended the process of smelting of ores (copper and iron) to account for the concentration of iron in the core of the Earth (Figs. 10A, B), when the slag (lighter part) gets separated from the metal (the heavier part). The Earth appears to have acquired its core of iron during the process of cooling of the primordial molten state.

What was the chemical composition of the entire Earth when it was in the (primordial) molten state? The average values of the contents of the elements of the crustal rocks upto a depth of 150 km are available (Table 5), but not those of the Core. Considering the superabundance and almost monopoly of Fe and Ni in the core, the hypothetical chemical composition of the entire Earth, has been projected in Table 15.

Table 15. Hypothetical chemical composition of the entire Earth

Elements	*Percentage*	*Elements in decreasing order of abundance*
Fe, Ni, Ti	60 to 70	Fe, Ni, Ti, Si, Al, Mg, Ca, Na, K
Si, Al, Mg, Ca, Na, K	30 to 40	

The hypothetical chemical composition of the entire Earth projects not only Fe to be the most abundant element, but even Ni and Ti to dominate over SiO_2, in the order of abundance.

CAUSE OF ABUNDANCE OF IRON IN THE EARTH

The cause could be derived from the process of formation of the elemtns, iron being one of them. According to Fowler (1967), all the elements are derived from the burning of hydrogen and helium at a very high temperature of 10-12 million degrees Kelvin, and that too within short span of time (fraction of second). The elements are then converted into stable forms. Some attain gaseous state (O, N, Ne), some combine with oxygen to form oxides (FeO, Fe_2O_3, TiO_2, SiO_2, H_2O etc.), while majority of them combine with SiO_2 to form silicates (felspars, pyroxenes etc.) Fowler's scheme (Fig. 1) indicates that the several elements are built through stages. As an example, the formation of the iron group and that of Na is given below :

(i) H $\rightarrow He^4 \rightarrow C^{12} \rightarrow O^{16} \rightarrow Ne^{20} \rightarrow$ Mg $\rightarrow$ Iron group

(ii) H $\rightarrow He^4 \rightarrow C^{12} \rightarrow O^{16} \rightarrow Ne^{20} \rightarrow Ne^{21} \rightarrow Ne^{22} \rightarrow$ Na (Sodium)

The act of development of the different elements from hydrogen passes through several intermediate stages. It is to be noted that the act has not only produced numerous elements, but some of them are greater in abundance. Thus O_2, N, Si, Fe are formed in a larger proportion, as can be judged from the fact :

(i) N and O_2 build up the extensive envelope of the atmosphere,
(ii) O_2 in combination with H has given rise to the Hydrosphere forming 3/4th surface of the Earth,
(iii) Si is the stem of all silicate minerals of the rocks building up the crust, and
(iv) Fe forming the core of the Earth having a radius of 5200 km.

Abundance of O_2, N, Si and Fe elements raises two questions, (i) how, (ii) what was the necessity. The abundance of H, O_2, N and He, has a philosophical background. Oxygen and Hydrogen have joined hands to form large volume of water which is essential to the animal and the plant life. Oxygen of the atmosphere keeps the human beings alive. The envelope of helium around the Earth is protecting the human life from the effects of the radiations emanating from the Sun. The Earth has its magnetic property due to the presence of iron in the core. The cause of abundance of the elements, therefore, goes beyond the materialistic view point and knocks at the door of philosophy. Is there any *remote control* of a *creator* in producing the said elements in abundance, or, is it to be categorised as an *accident* during the evolution of the Universe?

The modus operandi of creation of large quantities of a few elements needs an explanation. Fowler (1967) has putforth two ways in which iron is formed. These are as given below :

(i) H $\rightarrow He^4 \rightarrow C^{12} \rightarrow O^{16} \rightarrow Ne^{20} \rightarrow$ Mg $\rightarrow$ Iron group, and
(ii) H directly to iron group.

The cause of abundance of iron might be due to more time being spent over the conversion of Mg into iron [equation (i)], or iron was directly formed from hydrogen without passing through the intermediate stages as shown in equation (ii). However Si (Silicon) does not find any position in the Fowler's scheme of formation of the elements. Whatever could be its mode of evolution, Si is also required to be produced in large quantity, because it forms the largest constituent in the form of SiO_2 in the crustal rocks (Table 5).

Trendall (2000) has noted that :

(i) at 4.3 G.a., the Earth was totally molten with a differentiated core,

(ii) crust of the Earth was ultramafic between 4.3 and 4.0 G.a., which served as the source of iron for the much debated banded iron formation of the pre-Cambrian age, and

(iii) Earth's earliest atmosphere was rich in CO_2 which held the iron in ferrous state in the oceanic waters.

However, Trendall (op. cit.) has not described the composition of the core. The ultramafic rocks, however, are not extraordinarily rich in iron, the average content being 11.1 (Fe_2O_3 + FeO, Table VI, p. 20 of Tyrrell, 1922). These rocks, however, are richer in MgO (32.2, Table VI, p. 120 of Tyrrell, 1922) than in iron. What was the cause of richness of iron during 4.3-4.0 G.a. period?

From the foregoing account it sounds logical to infer that in the initial (primary) molten stage of the Earth, the heavier metalliferous elements (Fe, Ni, Ti, and the like) constituted as much as 60-70%, the remaining part comprising the lighter elements (Si, Al, Na, K etc.). Fe and Si dominated amongst the heavier and the lighter elements, respectively. Later, as the Earth started cooling, a large quantity of Fe and Ni formed the *core*, the rest of the Fe formed a constituent of the crustal rocks. Even then a million dollar question remains unanswered as to how Fe and Si were formed in greater quantity than the other elements that too during a short period of time of formation of all the elements from the burning of hydrogen and helium.

STATE OF MINERALS IN MAGMA

As already noted, the planets acquire their elemental composition at a very high temperature of 10-12 million degree Kelvin, and the process of formation of elements ceases at lower temperature. The quality and abundance of elements, thereafter, remains unchanged. The elements then combine to form various oxides, when the temperature is lowered to about 5000 degrees Kelvin. Therefore, the Earth too can be presumed to have acquired its chemical composition in a similar manner. The feature to be debated is whether the magma is composed of (a) oxides of several elements, or (b) the oxides of the elements are already bound up in several silicate minerals, which, however, are in a molten state. Dott and Batten (op. cit.) hold the view that "The most refractory elements Ca, Al and Ti condense and form chemical compounds in the range of 1500-1300°C, Fe, Ni, Co, Mg and Si condense and form chemical compounds in the range of 1300-1000°C. Remaining elements condense and form compounds between 1000-300°C. Water condenses below 100°C".

In any igneous body, the constituent minerals are more or less uniformly distributed. To elaborate this feature, in a granitic pluton, the K-felspars, the plagioclasses, the biotite and quartz, are seen to be evenly distributed. In other words it means that the individual minerals do not form clusters - only plagioclase, or quartz, or orthoclase, at several places within the granitic pluton. Chemical analysis of minerals reveals that these are made up of

SiO_2, Al_2O_3, CaO, MgO, FeO, Fe_2O_3, Na_2O, K_2O, TiO_2 in various proportions. Therefore, these oxides participate in the building up of the minerals from the magma. Some oxides like CaO is required in the formation of plagioclasses as well as the ferromagnesian minerals. The role of SiO_2 is still more complicated, because it constitutes the stem of every silicate mineral. It thus appears that the oxides of the elements migrate to various centres of crystallisation (of minerals) initiated in the (melt) magma.

A feature to regard the magma to be made up of oxides of various elements, is the occurrence of quartz in the acidic rocks like the granite, the granodiorite. If the magma was already made up of silicate minerals, oxide "SiO_2" will not be available to form grains of quartz. Likewise, veins of quartz can not be formed at all! The saturation principle of Shand, indicates that depending upon the availability of SiO_2 in the magma, it builds up oversaturated (Orthoclase, Albite), saturated (Pyroxenes, Amphiboles etc.) and unsaturated (Nepheline, Olivine, Felspathoids) minerals. The oxides of several elements can be expected in a primordial magma, which was formed at the time of the birth of the Earth. Later, however, when magma is regenerated through the melting of the crustal rocks, the various oxides of the elements are already bound up in the several minerals. Does the process of melting breaks up the minerals, and releases the oxides of the elements? Is the heat sufficient to bring about such a break up? The processes of magmatic differentiation and magmatic assimilation add to the complexity of the chemical status of the minerals (whether in the oxide or the silicate form) comprising magmas. Whether a magma is to be deemed as made up of (i) oxides of various elements, or (ii) silicates of various elements, certainly discerns attention of the petrologists.

CHAPTER 9

SOME FEATURES OF WATER AND GROUNDWATER

Amongst the several beneficial features of the Earth, water tops the list followed by oxygen. It is the grace of *God* that large volumes of water have been formed through the reaction between hydrogen and oxygen. The extent of hydrosphere is stupendous, and it forms 3/4th surface of the Earth. The initiation, survival and the growth of the animal and the plant life is unquestionably due to the presence of water and oxygen on the Earth. Water is an *Elixir* both to the animal and the plant life. It is an invaluable gift given by *Nature* to the mankind! Probably Earth is the only planet bestowed with water, as far as the Solar System is concerned.

NOMENCLATURE OF WATER

Water is given different names based on its composition, place of occurrence and utility. In terms of chemistry, it is described as an oxide of hydrogen, and is noted as H_2O. The name hydrogen is indeed thought provoking, because this element is regarded to "generate water". It is of interest to note that the prefix "hydro" stands for water, e.g., hydrostatic pressure, hydrochemistry, hydrology, hydrogeology and so on. The chemical compound H_2O is given the name *water*. The terms steam, water and ice, respectively stand for its gaseous, liquid and the solid states. The name "Heavy Water" which in chemical terminology is called as "Deuterium Water", is associated with the volcanic erruptions. Its formula is D_2O. The categorisation of water into "heavy and light" is done on the basis of the isotopic ratio of D and H. Heavy and light are the isotopic forms of water. The heavy water is named as Deuterium water, while the lighter one is the normal water composing the rain, rivers, groundwater and so on.

Water also is named according to its mode of occurrence. Water is all pervasive. Thus:

(a) it constitute a part of the atmosphere which is called as "moisture",

(b) clouds are entirely composed of water in the form of "vapour",

(c) being in liquid form at the ordinary temperature and pressure, it forms lakes, ponds, streams, rivers, seas and oceans. Water seeps into the crust and assumes the name of "groundwater or sub-surface water",

(d) it enters into the composition of some of the minerals that build up the crustal rocks, and

(e) volcanic erruptions emit large volumes of water besides other gases.

Question that crops up in mind is, whether water travels from the atmosphere (clouds) into the deeper parts of the Earth, or it travels upwards from depth to the surface, and thence into the atmosphere. The answer to it is tied up with the origin of water in the Earth.

The groundwater is technically known as "Meteoric water". Water trapped in the sediments during the process of deposition is called as "Connate". Water liberated during the igneous activity is given the name of "Juvenile". These names are thus based on the place of occurrence, or source of water. Is it possible that some water might have been entrapped at great depths, and it has not so far reached the surface of the Earth?

Let us now turn to the quality and appearance of water. A chemist describes it as a tasteless, odourless, colourless, transparent liquid. It is interesting to observe that the clouds which are composed of water vapour, are generally white, but those causing rain, are dark grey in colour. Two famous Indian rivers, Ganga and Yamuna present a spectacular sight at their confluence near Prayag, Allahabad district, U.P., India. The convergence is named as "Triveni Sangam". Water of Yamuna is brownish grey, while that of Ganga is white. It is a site of scenic beauty, and a scientific puzzle in respect of the contrast in colour of the water of the two rivers, though they have practically originated from one and the same place.

Theoretically purest water does not contain any dissolved matter in it. Purest water like the distilled water is tasteless. What then imparts taste to the "Potable Water"? In Table 16, the chemical composition of different categories of water is presented.

From Table 16 it is seen that the water primarily does not contain any cations in it, but these are later acquired by it from the atmosphere and the crustal rocks. The term "Fresh water" which is often used while describing the quality of groundwater, unintentionally conveys that it has been drawn afresh from a dug well or a borewell. The implication of that probably is that it is "compositionally fit" for human consumption. One another term in vogue is "potable water" which is a synonym for "fresh water". The taste of water is due to

Table 16. Chemical composition of different categories of water

Ingredient	Water					
	Distilled	Rain	Lake	River	Underground	Oceanic
Ca	Does not	0.50	13.00	32.00	300.00	79.00 NaCl
Na	contain any dissolved	0.20	75.00	190.00	450.00	$MgCl_2$ 16.00
Mg	salts	0.15	8.00	14.00	150.00	$MgSO_4$
K		–	–	–	–	4.00 $CaSO_4$
Cl		0.84	6.00	75.00	750.00	1.00 Others
HCO_3		1.47	196.00	255.00	250.00	100.00
SO_4		–	24.00	120.00	150.00	
NO_3		0.04	2.00	0.70	–	

Note : Contents of cations and anions are given in ppm, for the rain, lake, river and underground water. Contents of salts in the oceanic water are given in per cent. Oceanic water data taken from Jenson, M.L. and Bateman, A.M., p. 201, 1981.

the quality and the quantity of salts dissolved in it. The names "brackish, saline and brine" correspond to water containing 500 to 3300, 5000 to 15,000, more than 15,000 satts in ppm respectively. The general value of salinity of the sea water is 35,000 ppm. As the terms brackish and saline denote taste, the term "Sweet Water" be introduced to replace the term "fresh water" or "potable water".

CONTENTWISE DISTRIBUTION OF WATER

Water occurs in the form of oceans, ice caps, icebergs, glaciers, lakes, rivers, groundwater, and in the atmosphere. The oceanic water is saline, and therefore, it is not potable, the rest belongs to the category of sweet water. The non-potable oceanic water is 97.37 per cent in extent, leaving a mere 2.63 per cent to the potable or the sweet water category Out of the total volume of sweet water available in the Earth, 77.23 per cent of water is locked up in the ice caps, icebergs and the glaciers, 22.21 per cent constitutes the groundwater to a depth of 4000 meters, while lakes, rivers, soil moisture, atmosphere, taken together hold a mere 0.56 per cent of water. This statistics indicates acute shortage of sweet water on the Earth, and therefore, it needs conservation and a careful monitoring more seriously than that adopted for oil, natural gas and coal.

WHEN AND HOW OF FORMATION OF WATER, CAUSE OF SALINITY OF OCEANS

Chemically, water is an oxide of hydrogen. The elements involved are oxygen and hydrogen. All the elements are formed from the burning of hydrogen (Fig. 1), hydrogen itself being acquired from the Universe. The elements further combine with one another to build up several minerals - silicates, oxides, sulphides and so on. One of the outcome of such a process is water. It must have been in the gaseous state as long as the Earth was in the molten condition. In age, therefore, it should be deemed to be as old as the Earth - 4.5 b.y. The source of water for the formation of hydrosphere being the rains, the oceanic water must have been sweet to begin with. The water forming the rivers, the streams, the lakes, the ice-caps, the glaciers and the groundwater, is sweet. Though the rain water does not contain dissolved salts in it, the rivers acquire Na, Ca, Mg during their journey towards the oceans. The content of the dissolved salts (Na, Ca, Mg) is least in the rain water, low in the lakes, moderate in the rivers, and more in the groundwater (Table 16). However, the water of the said categories does not reach the category of "saline water". What then is the cause of very high contents of Na, Ca and Mg in the oceanic water?

Bateman, A.M. (1958) has categorised the quality of water on the values of salinity, and those are given in Table 17.

Table 17. Classification of quality of water (after Bateman, 1958, 2nd Edition, p. 866)

Contents of salts in ppm	Quality
400	No taste of salt
500	Slightly brackish taste
1000-2500	Strong taste but bearable
3300	Usable domestically
3500-5000	Almost unbearable
5000 ±	Unfit for human use
6250	Horses live on it in good condition
7800	Horses can live on it
9375	Cattle can live on it
15625	Sheep can live on it
16000	Beyond tolerance limit for most grasses
35000	Seawater

According to Jenson and Bateman (op. cit., p. 201), the total amount of salts in the ocean is 218 million cubic km. The contents of NaCl, $MgCl_2$ and $MgSO_4$, $CaSO_4$ and the other salts in per cent respectively are 79.0, 16.0, 4.0 and 1.0. The rivers contribute 4 billion tonnes of salts annually to the oceans. According to Jenson and Bateman (op. cit., p. 199), about 3.45 per cent of seawater consists of dissolved salts of which 99.75 per cent by weight is made up only of seven ions which are listed below.

Cations in per cent		*Anions in per cent*	
Na	30.61	Cl	55.04
Mg	3.69	SO_4	7.68

Ca	1.16	HCO_3	0.41
K	1.10		

From the foregoing account it is apparent that the content of the dissolved salts is extremely high in the oceanic water. Further Na predominates over Ca, Mg, while Cl dominates over HCO_3 and SO_4. The salinity is thus primarily due to NaCl i.e., the common salt. It is significant to note that the average content of Na_2O and CaO in the crustal rocks is 3.84 and 5.08, respectively (Table 5). But in the seawater, Na very much predominates over Ca. The content of the dissolved salts is more in the groundwater than that in the rivers and the lakes. This is so because the rain water acquires more quantity of cations (Na, Ca, Mg), while seeping through the disintegrated and decomposed crustal rocks. Chemically Na is more soluble than Ca and Mg, in the state of chloride. Further, out of the salts discharged by the rivers into the oceans, large quantities of Ca are consumed by the marine life in building up the shells, coral reefs and so on. Abstraction of Ca, therefore, relatively enriches the content of Na in the oceanic water. Thus the action of these processes over millions of years appears to be the cause of salinity and dominance of Na in the oceanic waters.

ABOUT THE MOVEMENT OF WATER AND GROUNDWATER TABLE

The first movement is vertically down in the form of the rains. If the Earth were not to provide any uneven surface, the rainwater will stop its further movement on reaching the surface of the land. The unevenness creates elevated landmasses, and water, therefore, will start moving down the slopes of the landmasses. This movement takes place till the water reaches the lakes, ponds, seas, and ultimately the oceans. The feature to be noted is that the water travels more diagonally (along the slopes of the ground) than vertically into the crustal rocks. If this be not true, then almost all of the rain water would have seeped into the crust, and resulted in much smaller sized, few seas and oceans. But the reality is contrary to it. The extent of the seas and the oceans is nearly 3/4th of the Earth's surface.

Rainwater seeping into the crustal rocks constitutes the groundwater. This seepage is in the vertical direction and is facilitated if the rocks are sheared, fractured and porous. Presence of an impervious horizon like clays, inhibits the downward movement. Further, rise of temperature with depth, prohibits water from seeping into the crust to deeper levels. The groundwater table is projected only for the continents, and that too upto reaching the coast. The trace of the watertable reveals that it is shallowest near the coast and it is deep below the continents. The shape of the topographical profile of most of the continents is such that, it is steep to begin with, moderate in the intermediate stretches, and extremely gentle in the proximity of the coast. It is because of the gentle topographical slope, the watertable is shallow near the coast. Thus the topographical slope of the ground controls the movement of the "run off" of rainwater in the diagonal or in the vertical direction, which in turn monitors the "Groundwater Table"

The foregoing account is a prelude to pose the question, "is there a watertable beneath the oeanic floor"? If yes, will the groundwater be potable or will it be extremely saline? The groundwater along the coastal track of any continent, however, is observed to be potable, though enormous volume of the oceanic water is in its closest vicinity. There are numerous instances of the occurrence of "potable water" on islands surrounded by the oceanic water. Some examples are : (a) Rameshwaram island, off the coast of Pamban, Tamil Nadu, India, (b) Sindhudurg island, Malvan, Maharashtra, India. These instances raise the probability of the existence of potable groundwater beneath the extensive oceanic floor. If this turns out to be a reality, could it be harnessed to meet the ever increasing demand of potable water on the continents?

The knowledge of (a) quality and quantity of groundwater, (b) the behaviour of the watertable of the continents and that of the oceanic crust, is of immense importance in the context of "exploration, exploitation and management" of providing potable water to the ever increasing human population, the world over. These and the other aspects of water mentioned in this chapter, certainly constitute matter to be dealt with, in a separate book, and not as a part of the present one, because the matter to be dealt with is varied and enormous.

REFERENCES

REFERENCES

Abel, G.O. (1969), Exploration of the Universe, 2nd Edn., Holt, Rinehart and Winston, Inc., New York.

Ahmed, F. (1961), Palaeogeography of the Gondwana period in Gondwanaland, with special reference to India and Australia and its bearing on the theory of Continental Drift, Mem. Geol. Surv. Ind., Vol. 90, p. 142.

Bateman, A.M. (1958), Economic Mineral Deposits, 2nd Edn., p. 866, John Wiley and Sons, New York.

Barth, T.F.W. (1962), Theoretical Petrology, 2nd Edn., Wiley, New York.

Bott, M.H.P. (1971), The Interior of the Earth, Edward Arnold, London.

Chamberlain, T.C. and Moulton, F.R. (1945), Planetesimal Hypothesis.

Cloud, P.E. (Jr.), (1968), Atmospheric and Hydrospheric Evolution on the Primitive Earth. Science, Vol. 160, pp. 729-736.

Condie, K.C. (1984), Plate Tectonics and Crustal Evolution, 2nd Edn., Pergamon Press.

Dietz, R.S. (1961), Continent and Ocean Basin Evolution by Spreading of the Sea Floor, Nature, Vol. 190, pp. 854-857.

De Buffon, G.L. (1949).

Deer, W.A., Howie, R.A. and Zussman, J. (1976), An Introduction to the Rock-Forming Minerals, 2nd Edn., ELBS and Longman.

Dott, R.H. and Batten, R.L. (1976), Evolution of Earth, McGraw-Hill Inc.

Du Toit, A.L. (1937), Our Wandering Continents, Oliver and Boyd, Edinburgh.

Dudley Stamp, L. (1938), A Concise Geography of the World, Longmans Green and Co. Ltd., p. 11.

Ferris, Timothy (1990), Coming of age in the Milkyway, Affiliated East-West Press Pvt. Ltd., New Delhi.

Fowler, W.A. (1967), Nuclear Astrophysics.

Fowler, W.A. (1984), Journal of Reviews of Modern Physics, Vol. 56, p. 149.

Gamow George (1961), The Creation of the Universe, Rev. Edn., The Viking Press Inc., New York.

Gamow George (1962), Gravity, Double Day Company Inc., Garden City, New York (Anchor Paper Back).

Gamow George (1963), A Plant called Earth. Viking Press, New York.

Gokhale, N.W. and Koppad, V.B. (1971), Xenolith bearing dyke of Belbatti, Shirbatti Taluka, Dharwad District, Mysore State, Kar. Univ. Jour. Sc., Vol. XVI, pp. 52-58.

Gokhale, N.W. and Puranik, S.C. (1981), On the occurrence of spherulitic structure in basic dyke from Doni areas, India. Acta. Geol. Acad. Sc. Hung., Vol. 24 (2-4), pp. 331-336.

Gokhale, N.W. (1996), Theory of Structural Geology. CBS Publishers and Distributors, New Delhi.

Gokhale, N.W. (2001), A Guide to Field Geology, pp. 51-52, Figs. 5a, 5b, CBS Publishers and Distributors, New Delhi.

Hartland, W.B. (1964), The Phanerozoic Time Scale, London Geological Society and 1971 Publication, No. 5.

Hawkes, H.E. and Webb, J.S. (1962), Geochemistry in Mineral Exploration, Harper and Row, New York, p. 415.

Hawking, S.W. (1989), A Brief History of Time, Bantam Books, New York.

Holland, T.H. (1941), The Evolution of Continents. A Possible Circulation of Conflicting Evidence. Proc. Royal Soc., Edinburgh, Vol. LXI, Part II, No. 13.

Holmes, A. (1960), Principles of Physical Geology, Thomas Nelson and Sons Ltd., 36, Park Street, London.

Holmes, A. (1966), Principles of Physical Geology, Thomas Nelson and Sons Ltd., 36, Park Street, London.

Holmes, A. (1981), Principles of Physical Geology, Thomas Nelson and Sons Ltd., 36, Park Street, London.

Holmes, A. and Dorris, L. (1978), Holmes Principles of Physical Geology, ELBS and Nelson.

Jacobs, J.A., Russell, R.D. and Wilson, J.T. (1971), Physics and Geology, McGraw-Hill, New York.

Jenson, M.L. and Bateman, A.M. (1981), Economic Mineral Deposits, John Wiley and Sons.

Karanth, K.R. (1987), Groundwater Assessment Development and Management, Tata McGraw-Hill Publishing Company Ltd.

Krane, K.S. (1988), Introduction to Nuclear Physics, Section 19.3, pp. 764-785, John Wiley and Sons, New York.

Kraus, E.H., Hunt, W.F. and Ramsdell, L.S. (1959), Mineralogy, 5th Edn., McGraw-Hill Company, New York.

Krishnan, M.S. (1968), Geology of India and Burma, Higginbothams (P) Ltd., Madras.

Kulp, T.L. (1961), Science V, 133, and Holland, W.B. (1971), The Phanerozoic Time Scale, Geological Society Publication, London, Vol. No. 5.

Lamb, S.H. (2002), Is it all in the Crust? Nature, Section, Earth Science, Vol. 420, pp. 130-131.

Lepp, H. (1973), Dynamic Earth, McGraw-Hill Inc., New York.

Malik, J.N., Sohoni, P.S., Karanth, R.V. and Mehr, S.S. (1999), Modern and Historic Seismicity of Kachch Peninsula, Western India. Jour. Geol. Soc. Ind. Vol. 54.

Mason, R.G. (1960), Geophysical Investigations of the Sea Floor. Liverpool and Manchester Geological Journal, Vol. 2, Part 3, pp. 389-410.

Mckenzie, D.P. (1970), Plate Tectonics and Continental Drift. Endeavour, Vol. 29, No. 106, pp. 39-44.

Oldham, T. (1883), Catalogue of Indian Earthquakes from the earliest times to the end of A.D., 1869, Geol. Surv. Ind. Mem. 19, pt. 3.

Oldham, R.D. (1900), Assam earthquake of 12th June 1897, Geol. Surv. Ind. Mem. 29.

Oldham, R.D. (1926), Cutch earthquake of 16th June, 1819, with a revision of the Great Earthquake of 12th June, 1897. Geol. Surv. Ind. Mem. 46, pt. 2.

Park, R.G. (1983), Foundation of Structural Geology, 2nd Edn., Blackie and Sons Ltd., USA.

Pendse, C.G. (1948), Earthquakes in India and Neighbourhood. India Met. Dept. Sci., Notes X, No. 129.

Perret, M.E. (1976), "The Origin of Oceans and Continents - a Complement to Wegener's theory", Geography of the Ocean. International Geography, Section 3, pp. 90-93. Distributed by Pergamon Press Ltd., Headington Hill Hall, Oxford, OX_3 OBW, England.

Radhakrishna, B.P. and Vaidyanadhan, R. (1997), Geology of Karnataka, 2nd Edn., Geol. Soc. Ind., p. 27.

Richter, C.F. (1969), Elementary Seismology, Eurasian Publishing House (Pvt.) Ltd., New Delhi.

Sagan, Carl (1985), Cosmos, Ballantine Books, New York.

Schlee, Susan (1975), A History of Oceanography, pp. 251, 317 to 362, Robert Dale and Company.

Shand, S.J. (1931), The study of rocks. Allen and Unwin (Murby) London.

Subramanian, V. (1979), Chemical and suspended chemical characteristics of River of India, Jour. Hydrology, Vol. 44, pp. 37-55.

Stears, J.A. (1964), Unstable Earth, Methuen, London.

Taylor, F.B. (1908), Bearing of the Tertiary Mountain Belt on the Origin of the Earth's Plan. Bull. Geol. Soc. Am. Vol. 21, pp. 179-226.

The Oxford School Atlas (1978), 24th Edition.

Trendall, A. (2000), The significance of Banded Iron Formation (BIF) in the pre-Cambrian Stratigraphic Record, Geoscientist, Vol. 10, No. 6.

Turekian, K.K. (1968), Oceans, Prentice Hall, Englewood Cliffs, N.J.

Tyrrell, G.W. (1929), The Principles of Petrology, Chapman and Hall Ltd., 11, New Fetter Lane, London.

Verhoogen, F.J., Turner, F.J., Weiss, L.E., Wahrhafting, C. and Fyfe, W.S. (1970), The Earth : An Introduction to Physical Geology, Holt, Rinehart and Winston, Inc., New York.

Water - Nature's Wonder, Hydrologic Cycle, Publication of National Institute of Hydrology, Roorkee.

Wegener, A. (1966), The Origin of Continents and Oceans, New York, Dover (Paper back, S., 1708, English translation of 4th Edn., 1920).

Wells, Spencer, The Journey of Man © Telegraph Group, London, London, News, 20.12.2002 (Noted by The Hindu, India's National Newspaper, 21.12.2002).

Williams, H.M. (1941), Crater Lake, The story of its Origin, University of California Press, Los Angeles.

Wilson, J.T. (1963), Continental Drift, Scientific American Vol. 208, pp. 86-100.

Wilson, J.T. (1965), A New Class of Faults and their bearing on Continental Drift, Nature, Vol. 207, pp. 343-347.